Kota Miura
Nataša Sladoje
Editors

Bioimage Data Analysis Workflows

Editors
Kota Miura
Im Neuenheimer Feld 267
Nikon Imaging Center Bioquant BQ 0004
Heidelberg, Germany

Nataša Sladoje
Department of Information Technology
Centre for Image Analysis,
Uppsala University
Uppsala, Sweden

https://doi.org/10.1007/978-3-030-22386-1

Preface

The often-posed question among life science researchers, "Which software tool is the best for bioimage analysis," indicates misunderstanding which calls for explanations. It appears that this question cannot be answered easily, maybe even not at all. Biological research problems are not general, and each of them questions specific events among various phenomena seen in biological systems. Therefore, the answer to the "Which is the best?" question to a high extent depends not only on the biological problem that is to be addressed but also on the specific goals and criteria to be met. Moreover, the misunderstanding seems to be based on an assumption that at some point in the future, there will be an almighty software tool for bioimage analysis that solves most of the problems just by clicking on a button. This, most likely, is simply a dream that may never come true.

Software tools are developed with their central value towards having generic applicability of offered functionality to be as wide as possible. In a sense, this is the agenda towards "almighty." On the other hand, the biological question asked by each researcher is unique and specific. "Novel findings," which biologists are seeking, come as answers to specific and original questions that others have not thought about or by using a novel method that others have not used to approach the mystery of biological systems. There is a clear gap between how bioimage analysis tools are developed and how biological questions are valued. The former is towards generality, and the latter is towards specificity.

The gap can be filled by designing unique combinations of general tools. More precisely, image analysis software tools should be used by a researcher in a one-and-only, specific way, by designing a customized workflow combining various suitable implementations of algorithms, to address a specific biological problem. Such novel designs help the researchers to see and quantify the biological system in a way that no one has done before. The highly desired optimal combination of the generality of the available software tools and the specificity of biological problem is thus achieved. The outcome can lead to outstanding scientific results. However, when this gap between generality and specificity is overlooked, bioimage analysis becomes simply a pain: life science researchers do not know how to approach it and benefit from it. Question such as "A great software tool is available in my computer but why can't I solve my problem?" can be rather frustrating.

As digital image data have become one of the fundamental infrastructures of biological research activities, students and researchers in the biomedical and life sciences more and more want to learn how to use the available tools. They want to know how to use various resources for image analysis and combine them to set up an appropriate workflow for addressing their own biological question. Getting used to bioimage analysis tools means learning about the various components that are available as a part of the software and becoming proficient in combining them for quantifying the biological systems.

The Network of European Bioimage Analysts (NEUBIAS) was established in 2016, with the aim to promote and share information about rich image analysis resources that have become widely available nowadays and to encourage, through education, their uses.

Nowadays, we can access many resources. These are good news, but at the same time, this variety of options may be overwhelming, leading to difficult choices regarding tools and resources most suitable for a particular problem and their most effective combinations for any specific purpose.

The aim of this textbook is to offer guidance in learning to make such choices. It provides "guided tours" through the five selected bioimage analysis workflows relevant in real biological studies, which combine different software packages and tools. Realistically, these workflows are not general and cannot be directly applied to other problems. However, the best (if not the only) way to learn to design own specialized workflows is to study the craft (approaches and solutions) of others. *Bioimage Data Analysis* (Wiley 2016) was published with the same motivation; this textbook is a sequel, contributing to the same goal. We hope to continue by including more bioimage analysis workflows and, by that, inspiring new creative solutions of life science problems.

One prominent contribution of the NEUBIAS team to the life science community is the conceptual apparatus required for swimming in the sea of rich image analysis resources: definitions of *components*, *collections*, and *workflows*. These notions are introduced and explained in ▶ Chap. 1 and then utilized in the subsequent ones.

▶ Chapter 2 focuses on a workflow for measuring the fluorescence intensity localized to the nuclear envelope. Automatic segmentation of the nuclear rim, based on thresholding and mathematical morphology, is iterated through multiple image frames to measure the changes in fluorescence intensity over time. ImageJ macro commands are recorded by the command recorder and converted to a stand-alone ImageJ macro.

▶ Chapter 3 offers a step-by-step guide through a procedure to build a macro for a 3D object-based colocalization, showing also how to extend and adjust the developed workflow to include intensity-based colocalization methods.

▶ Chapter 4 aims at teaching the principles and pitfalls of single particle tracking (SPT). Tracking is, in general, very important for dynamic studies; focus is on propagating object identities over time and subsequently computing relevant quantities from the identified tracks. The developed workflow combines tools available in ImageJ/Fiji (for generating the tracks) and in MATLAB (for analyzing them).

▶ Chapter 5 introduces some of the powerful and flexible image analysis methods native to MATLAB, also providing a crash course in programming for those with no, or limited, experience. The tools are used to simulate a time series of Brownian motion or diffusion process, to analyze time-series data, and to plot and export the results as figures ready for publication. The workflow presented in this chapter is quite powerful in analyzing tracking data such as those presented in ▶ Chap. 4.

▶ Chapter 6 presents the computational approach of registering images from different modalities based on manual selection of matching pairs of landmarks. The identification of sites of clathrin-mediated endocytosis by correlative light electron microscopy (CLEM) is used as an example on how to apply an image registration workflow based on MATLAB's image processing toolbox.

This textbook is the first bioimage analysis textbook published as an output of the common efforts of NEUBIAS, the Network of European Bioimage Analysts, funded under COST Action CA15124. We would like to thank the leaders of workgroups (WGs) in NEUBIAS: Sebastian Munck, Arne Seitz and Florian Levet (WG1 "Strategy"), Paula Sampaio and Irene Fondón (WG2 "Outreach"), Perrine Paul-Gilloteaux and Chong Zhang (WG4 "Webtool biii.eu"), Sébastien Tosi and Graeme Ball (WG5 "Benchmarking and Sample Datasets"), Julia Fernandez-Rodriguez and Clara Prats Gavalda (WG7 "Short-Term Scientific Missions and Career Path"), and Julien Colombelli (NEUBIAS Chair). Their efforts to create a synergistic effect of the diverse workgroup activities towards the establishment of "Bioimage Analysts" are the strong backbone that has led to the successful realization of this book. We are very much grateful to the reviewers of each chapter: Anna Klemm, Jan Eglinger, Marion Louveaux, Christian Tischer, and Ulrike Schulze. Their critical comments largely improved the presented workflows. We are particularly grateful to the authors of each workflow chapters: Fabrice P. Cordeliéres, Chong Zhang, Perrine Paul-Gilloteaux, Martin Schorb, Simon F. Nørrelykke, Jean-Yves Tinevez, and Sébastien Herbert. They have traveled together with selfless commitment to achieve the demanding publication format we chose, which is to offer both the normal printed textbook and the "continuously updated" online electronic version. The publication of this book was enabled by the financial support from the COST Association (funded through EU framework Horizon2020), through the granted project "A New Network of European Bioimage Analysts (NEUBIAS, COST Action CA15124)." Finally, we thank all the members of NEUBIAS who, with their enthusiasm and commitment to the network's activities, have contributed to keep the momentum of the initiative constantly high, a vital element to enable it to reach its objectives, including the publication of this book.

Nataša Sladoje
Uppsala, Sweden

Kota Miura
Heidelberg, Germany

Acknowledgements

This textbook is based upon the work from COST Action CA15124, supported by COST (European Cooperation in Science and Technology).

COST (European Cooperation in Science and Technology) is a funding agency for research and innovation networks. Our actions help connect research initiatives across Europe and enable scientists to grow their ideas by sharing them with their peers. This boosts their research, career, and innovation.

► www.cost.eu

Contents

Contributors

Julien Colombelli
Advanced Digital Microscopy Core Facility, Institute for Research in Biomedicine, IRB
Barcelona, Spain

Barcelona Institute of Science and Technology, BIST
Barcelona, Spain
julien.colombelli@irbbarcelona.org

Fabrice P. Cordelières
Bordeaux Imaging Center, UMS 3420
CNRS – Université de Bordeaux – US4 INSERM
Bordeaux, France

Pôle d'imagerie photonique, Centre Broca Nouvelle-Aquitaine
Bordeaux, France
fabrice.cordelieres@u-bordeaux.fr

Sébastien Herbert
Image Analysis Hub – C2RT – Institut Pasteur
Paris, France
sebastien.herbert@pasteur.fr

Kota Miura
Im Neuenheimer Feld 267
Nikon Imaging Center Bioquant BQ 0004
Heidelberg, Germany
miura@embl.de

Simon F. Nørrelykke
Image and Data Analysis Group, Scientific Center for Optical and Electron Microscopy, ETH
Zurich, Zurich, Switzerland
simon.noerrelykke@scopem.ethz.ch

Perrine Paul-Gilloteaux
SFR-Santé MicroPICell Facility, UNIV Nantes, INSERM, CNRS, CHU Nantes
Nantes, France

INSB France BioImaging
Nantes, France
Perrine.Paul-Gilloteaux@univ-nantes.fr

Martin Schorb
Electron Microscopy Core Facility, EMBL Heidelberg
Heidelberg, Germany
martin.schorb@embl.de

Jean-Yves Tinevez
Image Analysis Hub – C2RT – Institut Pasteur
Paris, France
tinevez@pasteur.fr

Sébastien Tosi
Advanced Digital Microscopy Core Facility, Institute for Research in Biomedicine, IRB
Barcelona, Spain

Barcelona Institute of Science and Technology, BIST
Barcelona, Spain
sebastien.tosi@irbbarcelona.org

Chong Zhang
SimBioSys Group, Pompeu Farba University
Barcelona, Spain
chong.zhang@upf.edu

Workflows and Components of Bioimage Analysis

Kota Miura, Perrine Paul-Gilloteaux, Sébastien Tosi, and Julien Colombelli

K. Miura, N. Sladoje (eds.), *Bioimage Data Analysis Workflows*, Learning Materials in Biosciences,
https://doi.org/10.1007/978-3-030-22386-1_1

1

What You Learn from This Chapter

Definitions of three types of bioimage analysis software—Component, Collection, and Workflow—are introduced in this chapter. The aim is to promote the structured designing of bioimage analysis methods, and to improve related learning and teaching.

1.1 Introduction

Software tools used for bioimage analysis tend to be seen as utilities that solve problems off-the-shelf. The extreme version of such is like: "If I know where to click, I can get good results!". In case of gaming software, as the user gets more used to the software, the user can achieve the final stage faster. To some extent, this might be true also with bioimage analysis software, but there is a big difference. As bioimage analysis is a part of scientific research, the goal to achieve is not to clear the common final stage that everyone heads toward, but something original that others have not found out. The difficulty of the usage of bioimage analysis software does not only reside in the hidden commands, but also in the fact that the user needs to come up with more-or-less original analysis. Then, how can we do something original using tools that are provided in public?

In this short chapter, we define several terms describing the world of bioimage analysis software, which are "workflows", "components", and "collections", and explain their relationships. We believe that clarifying the definition of these terms can contribute largely to those who want to learn bioimage analysis, as well as to those who need to design the teaching of bioimage analysis. The reason is that these terms link the generality of software packages provided in public, with the specificity and the originality of the analysis that one needs to achieve.

1.2 Types of Bioimage Analysis Software

Software packages such as ImageJ (Schneider et al. 2012),[1] MATLAB,[2] CellProfiler (Carpenter et al. 2006)[3] or ICY (de Chaumont et al. 2012)[4] are often used to analyze image data in life sciences. These software packages are "**collections**" of implementation of image processing and analysis algorithms. Libraries such as ImgLib2 (Pietzsch et al. 2012),[5] OpenCV (Bradski 2000),[6] ITK (Johnson et al. 2015a,b),[7] VTK (Schroeder et al. 2006),[8] and Scikit-Image (van der Walt et al. 2014)[9] are also packages of image processing and analysis algorithms, although with a different type of user interface that is not graphical. We invariably refer to them as "**collections**". To scientifically analyze and address an underlying biological problem, one needs to hand-pick some algorithms from these

1 ▶ https://imagej.org
2 ▶ https://nl.mathworks.com
3 ▶ https://cellprofiler.org/
4 ▶ http://icy.bioimageanalysis.org
5 ▶ https://imagej.net/ImgLib2
6 ▶ https://opencv.org
7 ▶ https://itk.org
8 ▶ https://vtk.org
9 ▶ https://scikit-image.org

collections, carefully adjust their functional parameters to the problem and assemble them in a meaningful order. Such a sequence of image processing algorithms with a specified parameter set is what we call a "**workflow**". The implementations of the algorithms that are used in the **workflows** are the "**components**" constituting that workflow (or "**workflow components**"). From the point of view of the expert who needs to assemble a workflow, a collection is a package bundling many different **components**. As an example, many plugins offered for ImageJ are mostly also **collections** (e.g. Trackmate (Tinevez et al. 2016),[10] 3D Suite (Ollion et al. 2013),[11] MosaicSuite[12]…), as they bundle multiple **components**. On the other hand, some plugins, such as Linear Kuwahara filter plugin,[13] are a single **component** implemented as a single plugin.

Each **workflow** is uniquely associated with a specific biological research project because the question asked therein as well as the acquired image quality are often unique. This calls for a unique combination of **components** and parameter set. Some **collections**, especially those designed with GUI, offer **workflow templates**. These templates are preassembled sequences of image processing tasks to solve a typical bioimage analysis problem; all one needs to do is to adjust the parameters of each step. For example, in the case of Trackmate plugin for ImageJ (Tinevez et al. 2016), a GUI wizard guides the user to choose an algorithm for each step among several candidates and also to adjust their parameters to achieve a successful particle tracking **workflow** (see ▶ Chap. 4). When these algorithms and parameters are set, the **workflow** is built. CellProfiler also has a helpful GUI that assists the user in building a **workflow** based on **workflow templates** (Carpenter et al. 2006). It allows the user to easily swap the algorithms for each step and test various parameter combinations. ◘ Figure 1.1 summarizes the above explanations.

Though such templates are available for some typical tasks, **collections** generally do not provide helpful clues to construct a **workflow**—choice of components to be used and approach taken to assemble those **components** depend on expert knowledge, empirical knowledge or testing. Since the biological questions are so diverse, the **workflow** often needs to be original and might not match any available **workflow templates**. Building a **workflow** from scratch needs some solid knowledge about the **components** and the ways to combine them. It also requires an understanding of the biological problem itself. Each **workflow** is in essence associated with a specific biological question, and this question together with the image acquisition setup affect the required precision of the analysis. For example, image data in general should not be analyzed at a precision higher than the physical resolution of the imaging system that captures those data.[14] In some cases, a higher precision does not imply more meaningful results just because such precision can be irrelevant to the biological question. These aspects should be carefully considered during the planning of the analysis and the choice of the **components**, together with the choice of statistical treatment.

Many biologists feel difficulty in analyzing image data, because of the lack in skills and knowledge to close the gap between a **collection** of **components** and a practical

---

10 ▶ https://imagej.net/TrackMate
11 ▶ http://imagejdocu.tudor.lu/doku.php?id=plugin:stacks:3d_ij_suite:start
12 ▶ http://mosaic.mpi-cbg.de/?q=downloads/imageJ
13 ▶ https://imagej.net/Linear_Kuwahara
14 If the model-based approach designed to compute sub-pixel resolution results is used e.g. single molecule localization microscopy, precision does go beyond the given optical resolution and the approach is thus validated.

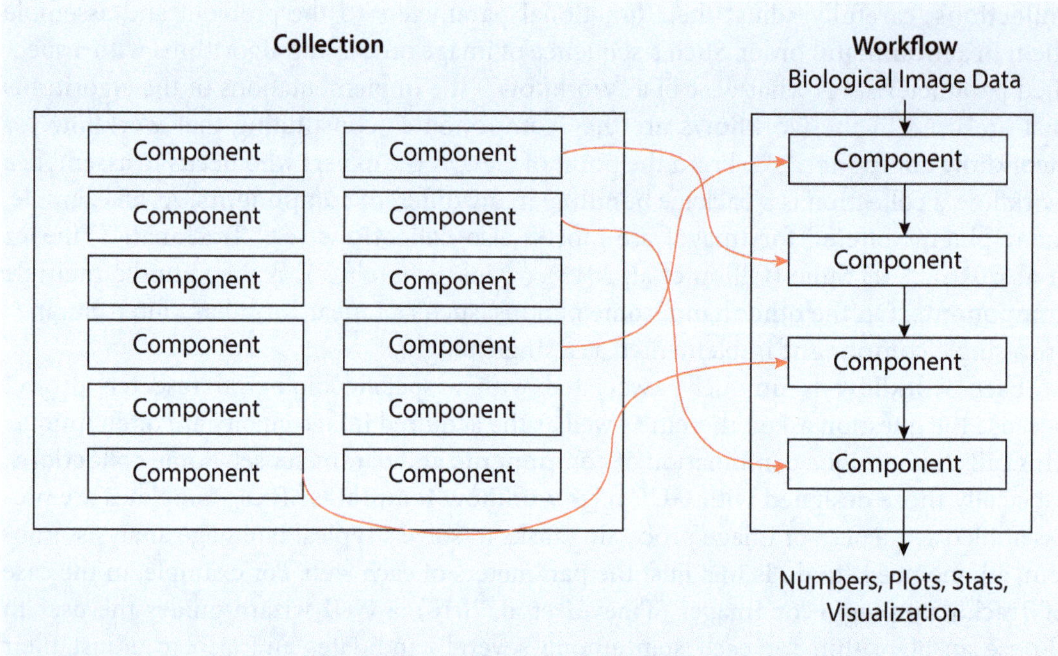

Fig. 1.1 Relationship between components, collection and workflow. Components (e.g. Gaussian blurring filter) are selected from collection (e.g. ImageJ) and assembled into a specific workflow (red arrow) for analyzing image data in each research project (e.g. scripts associated with journal papers)

workflow. A **collection** bundles **components** without **workflows**, but it is often erroneously assumed that installing a **collection** is enough for solving bioimage analysis problem. The truth is that expert knowledge is required to choose **components**, adjust their parameters and build a **workflow** (Fig. 1.1 red arrows). The correct assembly of **components** as an executable script is in general even more difficult, as it requires some programming skills. The use of **components** directly from library-type of **collections**, which host many useful **components**, also requires programming skills to access their API. Bioimage analysts may fill this gap but even they, who professionally analyze image data, need to always search for the most suitable **components** to solve problems, reaching the required accuracy or coping with huge data in a practical time.

Another important aspect and difficulty is the reproducibility of **workflows**. We often want to know how other people have performed image analysis and to learn from others new bioimage analysis strategies. In such cases, we look for **workflows** addressing a similar biological problem. However, many articles do not document the **workflows** they used in sufficient details to enable the reproducibility of the results. As an extreme example, we found articles with their image analysis description in Materials and Methods merely documenting that ImageJ was used for the image analysis. Such a minimalism should be strictly avoided. On the other hand, some **workflows** are written as a detailed text description in Materials and Methods sections in the publications. We go even further and recommend to publish **workflows** as executable scripts, i.e. a computer program, with documented parameter sets for clarity and reproducibility of analysis and results. In our opinion, the best format is a version-tracked script because the version used for the published results can be clearly stated and reused by others. A script embedded in a Docker image is even better for avoiding problems associated with a difference in execution environments.

Towards a more efficient designing of **workflows**, The Network of European Bioimage Analysts (NEUBIAS) has been developing a searchable index named Bioimage Informatics Search Engine (BISE). This service is accessible online at ▶ https://biii.eu and hosts the manually curated registry of **collections**, **workflows** and **components**.

Two ontologies are used for annotating resources registered to BISE: The BISE ontology for properties of resources e.g. programming language; and the EDAM Bioimaging Ontology (Kalaš et al. 2019)—an extension of the EDAM ontology (Ison et al. 2013) developed together with ELIXIR[15]—for applications of these resources, e.g. image processing step and imaging modality. "Component", "Workflow" and "Collections" are implemented as part of the BISE ontology for classifying the type of software, for more distinctive filtering of search results.

While BISE allows researchers to search for bioimage analysis resources at all these levels, general web search engines, such as Google, typically return hits of **collections** but not to the details of their **components**. In addition, **workflows** are in many cases hidden in biological papers and difficult to be discovered. BISE is also designed to feature users impressions on the usability of **components** and **workflows** so that individual experiences can be swiftly shared within the community.

Take Home Message

Within the world of bioimage analysis software, various types of tools, which can be classified as "collections", "components", or "workflows", coexist and are flatly provided to the public as "software tools". Clear definition of these types and recognition of the role of each is a foundation for learning and teaching bioimage analysis.

Further Readings

1. Miura and Tosi (2016) discusses the general challenges of bioimage analysis.
2. Miura and Tosi (2017) provides more details on the structure and designing of bioimage analysis workflows.
3. Details about NEUBIAS can be found at the following web pages:
   - ▶ http://neubias.org
   - ▶ https://www.cost.eu/actions/CA15124: The Memorandum of Understanding describes the objectives of the network, that includes the motivation to create the registry ▶ http://biii.eu.

Acknowledgements We are grateful to Nataša Sladoje for critically reading this text. We thank Matúš Kalaš for checking the text and correcting our mistakes.

15 ▶ https://www.elixir-europe.org

1

Bibliography

Bradski G (2000) The OpenCV Library. Dr. Dobb's Journal of Software Tools. http://www.drdobbs.com/open-source/the-opencv-library/184404319

Carpenter AE, Jones TR, Lamprecht MR, Clarke C, Kang IH, Friman O, Guertin DA, Chang JH, Lindquist RA, Moffat J, Golland P, Sabatini DM (2006) CellProfiler: image analysis software for identifying and quantifying cell phenotypes. Genome Biol 7(10):R100. ISSN: 1465-6914. https://doi.org/10.1186/gb-2006-7-10-r100

de Chaumont F, Dallongeville S, Chenouard N, Hervé N, Pop S, Provoost T, Meas-Yedid V, Pankajakshan P, Lecomte T, Le Montagner Y, Lagache T, Dufour A, Olivo-Marin J-C (2012) Icy: an open bioimage informatics platform for extended reproducible research. ISSN: 1548-7091

Ison J, Kalaš M, Jonassen I, Bolser D, Uludag M, McWilliam H, Malone J, Lopez R, Pettifer S, Rice P (2013) EDAM: an ontology of bioinformatics operations, types of data and identifiers, topics, and formats. Bioinformatics 29(10):1325–1332. https://doi.org/10.1093/bioinformatics/btt113

Johnson HJ, McCormick MM, Ibanez L (2015a) Template: the ITK software guide book 1: introduction and development guidelines, vol 1

Johnson HJ, McCormick MM, Ibanez L (2015b) Template: the ITK software guide book 2: design and functionality, vol 2

Kalaš M, Plantard L, Sladoje N, Lindblad J, Kirschmann MA, Jones M, Chessel A, Scholz LA, Rössler F, Dufour A, Bogovic J, Waithe CZD, Sampaio P, Paavolainen L, Hörl D, Munck S, Golani O, Moore J, Gaignardand A, Levet F, Participants in the NEUBIAS Taggathons, Paul-Gilloteaux P, Ison J, the EDAM Dev Team, Miura K, Colombelli J, Welcoming New Contributors (2019) EDAM-bioimaging: the ontology of bioimage informatics operations, topics, data, and formats (2019 update) [version 1; not peer reviewed]. F1000Research 8(ELIXIR):158. https://doi.org/10.7490/f1000research.1116432.1

Miura K, Tosi S (2016) Introduction. In: Miura K (ed) Bioimage data analysis. Wiley-VCH, Weinheim, pp 1–3. ISBN: 978-3-527-34122-1

Miura K, Tosi S (2017) Epilogue: a framework for bioimage analysis. In: Standard and super-resolution bioimaging data analysis. Wiley, Chichester, pp 269–284. https://doi.org/10.1002/9781119096948.ch11

Ollion J, Cochennec J, Loll F, Escudé C, Boudier T (2013) TANGO: a generic tool for high-throughput 3D image analysis for studying nuclear organization. Bioinformatics 29(14):1840–1841. ISSN: 13674803. https://doi.org/10.1093/bioinformatics/btt276

Pietzsch T, Preibisch S, Tomančák P, Saalfeld S (2012) ImgLib2–generic image processing in Java. Bioinformatics (Oxford, England) 28(22):3009–3011. ISSN: 1367-4811. https://doi.org/10.1093/bioinformatics/bts543

Schneider CA, Rasband WS, Eliceiri KW (2002) NIH Image to ImageJ: 25 years of image analysis. Nat Methods 9(7):671–675 (2012). ISSN: 1548-7105. http://www.ncbi.nlm.nih.gov/pubmed/22930834

Schroeder W, Martin K, Lorensen B (2006) The visualization toolkit, 4th edn. Kitware Inc. ISBN: 978-1930934191. https://vtk.org/

Tinevez J-Y, Perry N, Schindelin J, Hoopes GM, Reynolds GD, Laplantine E, Bednarek SY, Shorte SL, Eliceiri KW (2016) TrackMate: an open and extensible platform for single-particle tracking. Methods. ISSN: 10462023. https://doi.org/10.1016/j.ymeth.2016.09.016. http://linkinghub.elsevier.com/retrieve/pii/S1046202316303346

van der Walt S, Schönberger JL, Nunez-Iglesias J, Boulogne F, Warner JD, Yager N, Gouillart E, Yu T (2014) Scikit-image: image processing in Python. PeerJ 2:e453. ISSN: 2167-8359. https://doi.org/10.7717/peerj.453. https://peerj.com/articles/453

Measurements of Intensity Dynamics at the Periphery of the Nucleus

Kota Miura

K. Miura, N. Sladoje (eds.), *Bioimage Data Analysis Workflows*, Learning Materials in Biosciences,
https://doi.org/10.1007/978-3-030-22386-1_2

What You Learn from This Chapter

The aim of this chapter is to learn how to construct a workflow for measuring the fluorescence intensity localized to the nuclear envelope. For this purpose, the nucleus image is segmented to create a mask along the nuclear rim. The reader will learn a typical technique for automatically delineating the segmented area by post-processing using the mathematical morphology algorithm, and how to loop that piece of ImageJ macro and iterate through multiple image frames to measure changes in fluorescence intensity over time. This chapter is also a good guide for learning how to convert ImageJ macro commands recorded by the Command Recorder to a stand-alone ImageJ macro.

2.1 Introduction

In some biological research projects, we encounter problems that should be studied by measuring fluorescence intensity at the boundary between two different compartments. Here, we pick up an example analysis of the Lamin B receptor protein density targeting inner nuclear membrane. The protein changes its location from the cytoplasmic area (Endoplasmic Reticulum, ER) to the nuclear envelope (Boni et al. 2015).

We analyze a two-channel time-lapse image stack, a sequence of the process of the protein re-localization that causes increases in the protein density at the nuclear envelope. The data was acquired by Andreas Boni (Jan Ellenberg lab, EMBL Heidelberg) and have been used in many training workshops in EMBL as a great example for learning bioimage analysis. His work, with more advanced bioimage analysis workflows for analyzing the protein targeting dynamics, is published in The Journal of Cell Biology (Boni et al. 2015). Those codes and image data used in his study, which might be interesting for you after going through this chapter, are accessible through the supplementary data section in the journal website.[1]

Two images shown in ◘ Fig. 2.1 are from the first and the last time points of a time-lapse sequence.[2] Compare these images carefully. The green signal broadly distributed in the cytoplasmic area at time point 1 becomes accumulated at the periphery of nuclei (red) at time point 15—between these image frames, the signal changed its localization from ER to the nuclear envelope. We construct a workflow that measures this accumulation process by writing an ImageJ macro. The workflow involves two steps: First, we segment the rim of nucleus—nuclear membrane—using the first channel (histone). Second, we use that segmented nuclear rim as a mask to measure the intensity changes over time in the second channel.

Segmentation of nucleus using its marker (e.g. DAPI) is a popular image analysis technique used in many biological research projects, but to measure more specific location—in our case nuclear envelope—we need to add several more steps to refine the region-of-interest. When we are successful in determining the area of nuclear envelope, the measurement of intensity in that region over time is rather trivial. We just need to loop the same process for each time point. Especially for the analysis of time-lapse sequence, programming is highly recommended to iterate the measurement for each time point.

1 ▸ http://jcb.rupress.org/content/209/5/705
2 The images shown in the ◘ Fig. 2.1 are from a 4D hyperstack "NPC1.tif", which can be downloaded using ImageJ plugin "CMCI-EMBL". More details are in "Dataset" section.

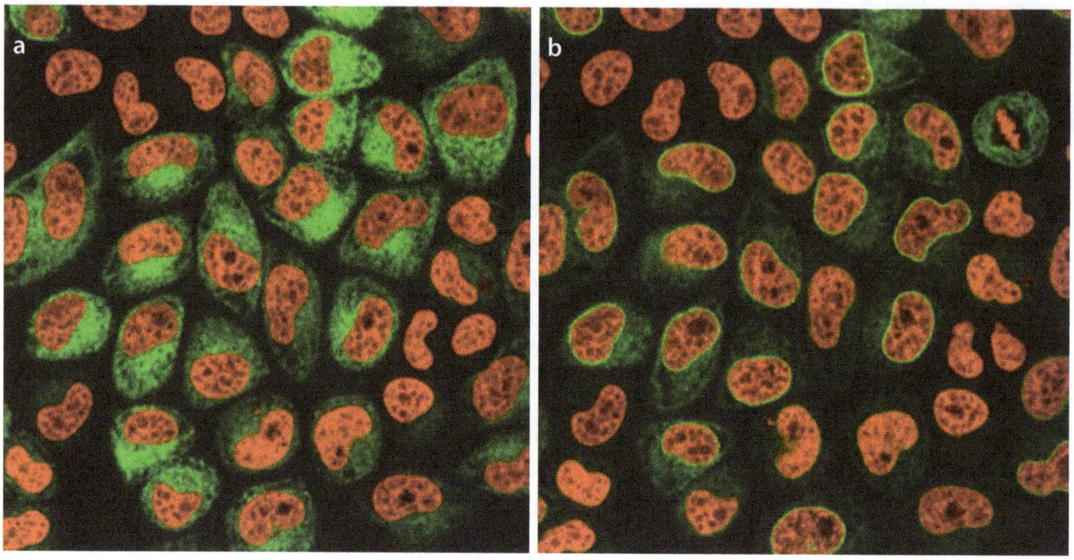

Fig. 2.1 Lamin receptor localization difference at two time points: More Lamin receptor in nucleus periphery. **a** Time point 1. **b** Time point 15

This chapter should be a good guide not only limited to study the intensity changes occurring at the nuclear envelope, but also in general for segmenting the edge (perimeter) of biological compartments such as the edge of organelle, plasma membrane and tissue boundaries. In principle, similar post-processing strategy is also applicable to 3D volumes by using 3D morphology filters.

2.2 Tools

We use Fiji (Fiji Is Just ImageJ) for image analysis.

- Fiji
  - Download URL: ▶ https://imagej.net/Fiji/Downloads
  - Please choose the latest version.

In addition, a plugin is required for loading the sample image data. Using the "Update sites" function, please add "CMCI-EMBL" to your Fiji installation. Please restart Fiji after this plugin installation.

2.3 Dataset

All ImageJ macro codes can be downloaded from the Github repository.[3]

The image data we used in this chapter can be downloaded using the plugin "CMCI-EMBL". After installation of this plugin, select the menu item `[EMBL > Sample Images > NPCsingleNucleus.tif]` to load the image data. This is a time-lapse

3 ▶ https://github.com/miura/NucleusRimIntensityMeasurementsV2/

sequence of a cell, extracted from "NPC1.tif" which can be also downloaded through the same plugin.

- Cell Type: Hela Cells
- Scale: 0.165 µm/pixel
- Frame Rate: 400 Sec/Frame
- Channels
  - Red channel (C1): H2B-mCherry (ex:561nm)
  - Green Channel (C2): Lamin B Receptor-GFP (ex:488nm)

2.4 Workflow

To simplify the development, we focus on a single cell/nucleus to construct the workflow. Load the image stack **NPCsingleNucleus.tif**. This is a hyperstack sequence. Slide the scroll bar at the bottom back-and-forth to watch the process of intensity changes. H2B-mCherry signal (red), used as a marker for nucleus, is more or less constant with its distribution. On the other hand, the Lamin receptor signal (green) exhibits strong accumulation to the nuclear membrane. To study this accumulation process, our aim is to measure the intensity changes of green signal intensity at the rim of the nucleus over time. The outline of the workflow is shown in the diagram (Fig. 2.2).

To achieve this aim we first need to identify the region of nucleus rim ("segmentation")—in other words, we create a mask of the nucleus rim. Using this mask we measure the changes in intensity over time.

2.4.1 Segmentation of Nucleus Rim

We first write a macro for the nucleus rim segmentation by taking following steps:

1. Split the original multi-channel image stack and create two image stacks of each channel for processing them independently (Fig. 2.3a)
2. Blur the image to attenuate noise (Fig. 2.3b)
3. Nucleus segmentation: Binarize the image by intensity thresholding (Fig. 2.3c)
4. Remove other Nuclei: At the right-bottom corner of the image, a small part of different nucleus is present. This should be removed.
5. Duplicate the image
   (a) Erode the original (Fig. 2.3e)
   (b) Dilate the duplicated (Fig. 2.3d)
6. Subtract the eroded from the dilated (Fig. 2.3f)

In the following we record these steps as macro commands using the Command Recorder (`[Plugins > Macros > Record...]`). We recommend you NOT to launch the command recorder from the beginning. Please first try to reproduce the workflow using mouse and the graphical user interface (GUI). This is like a rehearsal before recording your actions. When you become clear with the steps you have to take, record the processing steps. When you use the command recorder, be sure that "Macro" is selected in the "Record:" drop down menu at the top-left corner of the recorder.

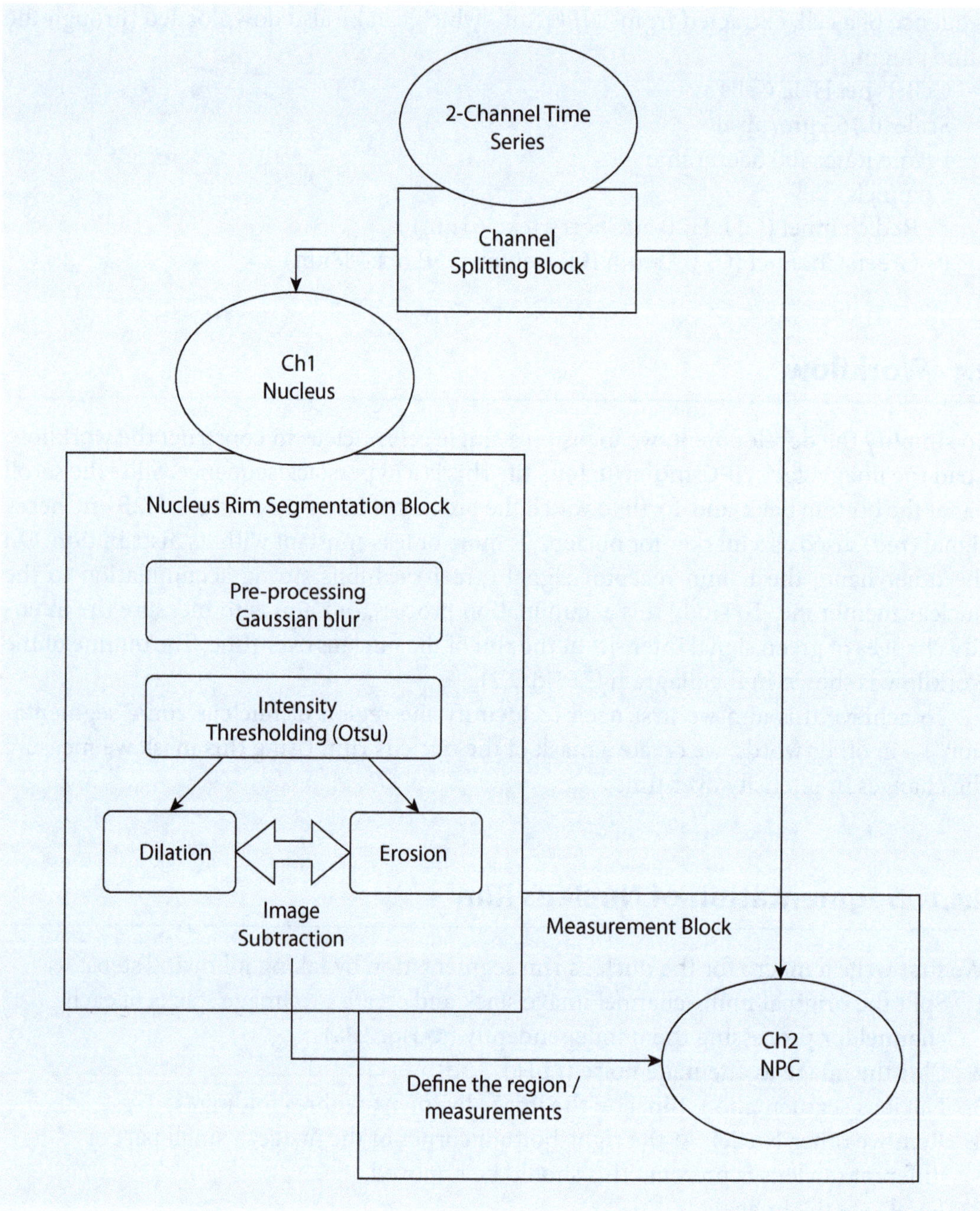

Fig. 2.2 The outline of the workflow

2.4.1.1 Block 1: Splitting Channels

To split the multichannel image stack from the GUI menu, do `[Image > Color > Split Channels]`. In the Recorder you will see the following command.

```
run("Split Channels");
```

`run` function is the most frequently used build-in macro function.

2

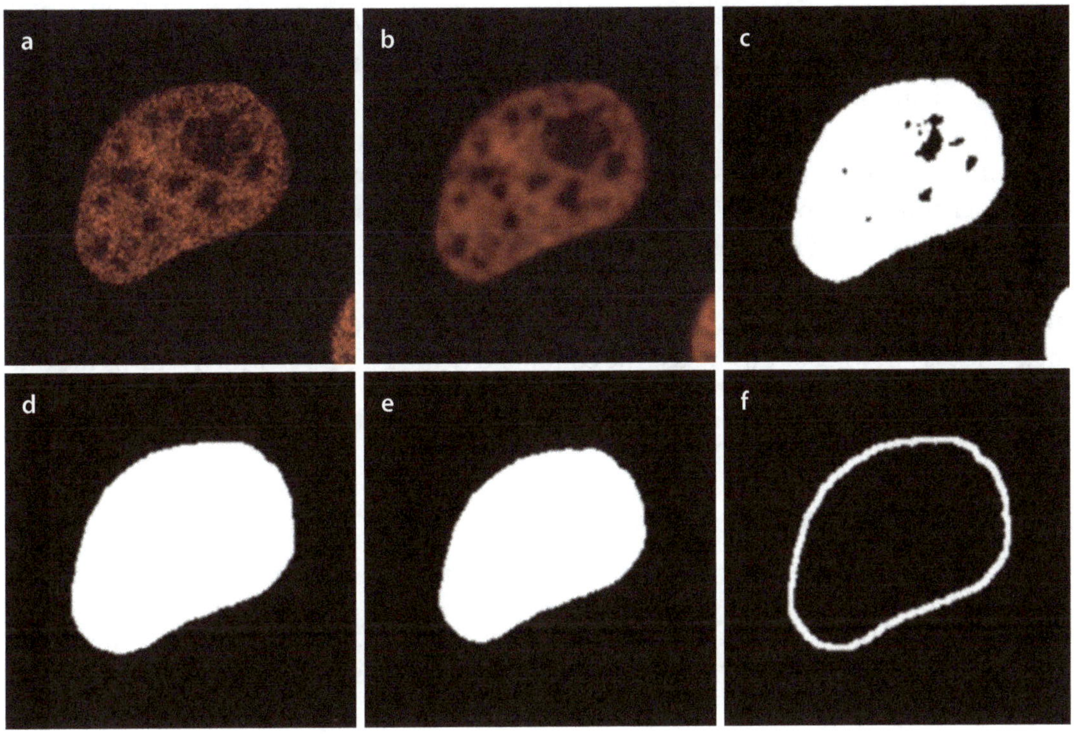

Fig. 2.3 The strategy of segmentation. **a** Original nucleus image. **b** Blurred nucleus image. **c** Binarized image, after thresholding. **d** Dilated binary image. **e** Eroded binary image. **f** Subtraction result, the rim

run("command"[, "options"])

Executes an ImageJ menu command. The optional second argument contains values that are automatically entered into dialog boxes (must be GenericDialog or OpenDialog). Use the Command Recorder (Plugins>Macros>Record) to generate run() function calls. Use string concatenation to pass a variable as an argument. With ImageJ 1.43 and later, variables can be passed without using string concatenation by adding "&" to the variable name.

The run function takes a menu item as the first argument and optional values (values you fill-in in a dialog window) in the second argument. In case of channel splitting, there is no such optional value so the second argument is ignored.

We then process the nucleus image. Click the nucleus image window to bring it up to the top—We call this action as "activating a window". By this clicking, we activated Channel 2 (red, nucleus image).

Please confirm that a new command shown below, is added to the recorder after activating the nucleus image.

```
selectWindow("C1-NPCsingleNucleus.tif");
```

...Here is the explanation from the macro function reference.

selectWindow("name")

Activates the window with the title "name".

This function takes a window title as an argument and activates a window with that title. When we used mouse to activate the nucleus channel window, we did it manually by visually recognizing the red nucleus image. On the other hand, in macro, we need to know the title of the windows of each individual channels to activate a specific window to provide it to macro as an argument of "selectWindow" command. How can we get the name of the nucleus channel window after splitting the channels of the original image?

Standard behavior of "Split Channel" command is that it automatically names the resulting stacks of individual channels by prefixing "C1-" or "C2-" or "C3" to the original image title. Based on this known behavior, we can construct these names if the original image title is known. For this we use the command `getTitle( )` which returns the title of currently active window as a string.

getTitle()

Returns the title of the current image.

Here is the code to activate the nucleus channel automatically after the splitting. More importantly, we also acquire "image ID". This will be explained later.

code/code_block1_ChannelSplitting.ijm

```
1 orgName = getTitle();
2 run("Split Channels");
3 c1name = "C1-" + orgName;
4 c2name = "C2-" + orgName;
5 selectWindow(c1name);
6 c1id = getImageID();
7 selectWindow(c2name);
8 c2id = getImageID();
```

Details:

- The first line grabs the window title as a string variable "orgName".
- The second line splits the stacks to each individual channel stack.
- 3rd and 4th lines compose the window title for each channel stack.
- 5th line activates the channel 1 stack.
- 6th line acquires the image ID of channel 1 stack.
- 7th line activates the channel 2 stack.
- 8th line acquires the image ID of channel 2 stack.

In lines 6 and 8, we acquire **image IDs**. Here is some more explanation about this: Each window has a unique ID number. To get this ID number from each image we use the command `getImageID( )`.

getImageID()

Returns the unique ID (a negative number) of the active image. Use the selectImage(id), isOpen(id) and isActive(id) functions to activate an image or to determine if it is open or active.

A window can be activated by `selectWindow` using its window title, but this could have a problem if there is another window with same name. Image ID has less problem since it is uniquely given to each window. To select a window using image ID, we use `selectImage(ID)` command.

> **selectImage(id)**
>
> Activates the image with the specified ID (a negative number). If id is greater than zero, activates the ID-th image listed in the Window menu. The ID can also be an image title (a string).

We acquire image IDs just after the splitting. From here on, we will use image IDs when we want to specify the image window we want to work on and to activate it.

Exercise 2.1

Test the code below and run it on several image windows. Confirm that each window has an unique ID number. Please ignore the line numbers when you write the code.

```
id = getImageID();
print(id);
name = getTitle();
print(name);
```

Save the channel splitting macro. When you name the file, add an extension ".ijm", as this indicates that the file is an ImageJ macro. This is only a part of the final workflow, and we call such part as a **"block"** of the workflow, and by assembling blocks with various functions, we construct a **workflow**. A block is a functional unit within the workflow. Each block is consisting of several **components**, each of which is the build-in function that implements a certain algorithm (see ▶ Chap. 1).

In the current case, we just finished the **Channel Splitting Block**, consisting of a channel splitter component, a window title getter component, an image window ID getter component, and window selector components.

2.4.1.2 Block 2: Segmentation of Nucleus Rim

Now we start working on the segmentation of nucleus rim. For this, we use only the nucleus image stack (channel 1) we got in the Block 1. Create a new tab in the script editor by `[File > New]`. We use this blank editor to write the next **block** for the detection of nucleus rim. We assemble all blocks as a single workflow later.

Following is the step-by-step procedure. Try first using the GUI (your mouse and the menu bar!). Then launch the Command recorder, redo the steps to record the history of commands. I recommend you to do so mainly because the initial trials with GUI let you visually understand what is going on, and also to get used to the sequence of operation for the command recording.

1. Gaussian Blur
   - `[Process > Filter > Gaussian Blur]`, sigma = 1.5, tick "Do Stack"
   - This diminutive blurring of the image attenuates noise and allows a better result for the segmentation.

2. Find Threshold
   - `[Image > Adjust > Threshold]`, select Otsu method
   - This simply changes the LUT, but not the data
3. Apply Threshold: Click 'Apply'
   - Changes the data to black and white using the threshold value using the Otsu method.
4. Find Threshold again (Otsu method)
   - We do this again for selecting the nucleus for the "AnalyzeParticle" in the following step.
5. Analyze Particles
   - `[Analyze > Analyze Particles]`
   - Options:
     - Size: 800-Infinity
     - Tick "Pixel Units"
     - Circularity: default (0–1.0)
     - Show: Mask
     - Tick Display Results, Clear results, Exclude on edges, Include holes.
   - We use AnalyzeParticle as a filter for segmented object. In our case, this filtering removes nucleus touching the edge of image. This way of usage is also effective in removing small none-nucleus signals.
6. Invert the LUT of the "Mask" created by AnalyzeParticle, so operations to be done in the following recognizes nucleus as the target of Dilation and Erosion.
   - `[Image > Look-up Table > Invert LUT]`
7. Duplicate the "Mask" Stack, and then apply "Dilation" to the original stack and apply "Erosion" to the duplicated.
   - `[Image > Duplicate]`
   - Set Iterations `[Process > Binary > Options]`
     - iterations 2 or 3
     - Tick dark background
   - Original: Dilate `[Process > Binary > Dilate]`
     - This increases the edge of nucleus by 2 or 3 pixels.
   - Duplicate: Erode `[Process > Binary > Erode]`
     - This decreases the edge of nucleus by 2 or 3 pixels.
8. Image Subtraction
   - `[Process > Image Calculator]`
   - tick "keep original", compute the difference of Dilated and Eroded.
     - Result: a band of 4 or 6 pixels at the edge of nucleus.

When you are done with the macro recording, check the results in the recorder. Below is an example of the output from the recorder.

code/code_block2_recordNucSeg.ijm

```
selectWindow("C1-NPCsingleNucleus.tif");
run("Gaussian Blur...", "sigma=1.50 stack");
 //run("Threshold...");
setAutoThreshold("Otsu dark");
```

```
6   setOption("BlackBackground", true);
7   run("Convert to Mask", "method=Otsu background=Dark calculate
    black");
8   //run("Threshold...");
9   run("Analyze Particles...", "size=800-Infinity pixel circular-
    ity=0.00-1.00 show=Masks display exclude clear include stack");
10  run("Invert LUT");
11  run("Duplicate...", "title=[Mask of C1-NPCsingleNucleus-1.tif]
    duplicate range=1-15");
12  selectWindow("Mask of C1-NPCsingleNucleus.tif");
13  run("Options...", "iterations=2 count=1 black edm=Overwrite
    do=Nothing");
14  run("Dilate", "stack");
15  selectWindow("Mask of C1-NPCsingleNucleus-1.tif");
16  run("Erode", "stack");
17  imageCalculator("Difference create stack", "Mask of C1-
    NPCsingleNucleus.tif", "Mask of C1-NPCsingleNucleus-1.tif");
18  selectWindow("Result of Mask of C1-NPCsingleNucleus.tif");
```

This recorded macro already runs properly as it is, but there is a problem: the code works only with image data with a specific window title. See the line 1. The command looks like this.

`selectWindow("C1-NPCsingleNucleus.tif");`

The window title given in the argument of `selectWindow` is hard-coded, so that if you need to apply this macro to a image data with a different window title, it will not work. The macro needs to be improved to allow the general applicability to other images.

For this reason, we need to change the code so that it uses ImageID instead of a fixed image title. Since the ImageID of the nucleus channel was already acquired after splitting the original image, we can use that ID to activate a specific image window.

As we are working separately from the channel splitting block, we assume that the nucleus channel stack is active and is the top window at the starting of current code. We replace the first line `selectWindow` with `getImageID( )` command to capture the ID number of the nucleus image window. Next, we need to add `getImageID` in line 10 and 13 to capture IDs of newly created windows. Due to these changes, we need to replace `selectWindow` in line 12 and 15 to `selectImage` to consistently use ImageID for accessing specific window. After these replacement, the updated code will look like the one shown below.

code/code_block2_recordNucSegV2.ijm

```
1   orgID = getImageID();
2   run("Gaussian Blur...", "sigma=1.50 stack");
3
4   //run("Threshold...");
5   setAutoThreshold("Otsu dark");
6   setOption("BlackBackground", true);
7   run(Convert to Mask", "method=Otsu background=Dark calculate
    black");
8   //run("Threshold...");
9   run("Analyze Particles...", "size=800-Infinity pixel circular-
    ity=0.00-1.00 show=Masks display exclude clear include stack");
10  dilateID = getImageID();
```

```
11  run("Invert LUT");
12  run("Duplicate...", "title=[Mask of C1-NPCsingleNucleus-1.tif]
    duplicate range=1-15");
13  erodeID = getImageID();
14  //selectWindow("Mask of C1-NPCsingleNucleus.tif");
15  selectImage(duplicateID);
16  run("Options...", "iterations=2 count=1 black edm=Overwrite
    do=Nothing");
17  run("Dilate", "stack");
18  //selectWindow("Mask of C1-NPCsingleNucleus-1.tif");
19  selectImage(erodeID);
20  run("Erode", "stack");
21  //imageCalculator("Difference create stack", "Mask of C1-
    NPCsingleNucleus.tif","Mask of C1-NPCsingleNucleus-1.tif");
22  imageCalculator("Difference create stack", dilateID, erodeID);
```

Here is the explanation of what was done.

- line 1: The first line is replaced with the `getImageID( )` command.
- line 10: `getImageID( )` command was inserted for a new image created by Analyze Particle command (in line 9). The new image is the mask that is eliminated with edge-touching nucleus.
- line 13: `getImageID( )` command was inserted for the duplicated image.
- line 15: The `selectWindow` command in line 14 was commented out and replaced by the `selectImage` command.
- line 19: `selectWindow` command is replaced by the `selectImage` command.
- line 22: Because we now have ImageIDs of both dilated and eroded images, we replace the titles of image windows with imageIDs for image calculator arguments. Compare the line 21 (commented out) and the line 22.

We are now almost done with the generalization of the nucleus rim segmentation block, but there still is a part that can be more general instead of a fixed window name. See line 12. This line uses `run` command to duplicate the "Mask" stack.

```
run("Duplicate...", "title=[Mask of C1-NPCsingleNucleus-1.tif] duplicate range=1-15");
```

The first argument "Duplicate..." is the name of the menu item `[Image > Duplicate...]` and this is OK.

The second argument contains multiple optional values you chose in the GUI. The first is the title of the duplicated image, that was automatically created by suffixing "-1" to the image title. Square brackets surrounding this new image title is for avoiding the problem with spaces in the image title, because spaces are used as the separator for the options in the second argument. **`duplicate`** is a keyword of a checkbox in the duplication dialog, for choosing whether to duplicate multiple frames in a stack or just a single currently shown frame. The third option is the frame range (`range=`), which defines the range of frames to be duplicated. Since we want to duplicate all frames, the range is set to `1-15`, from the first frame to the last 15th frame.

Within this second argument, two values in this command are not flexible enough for applying the macro to other images with different names. First is the image title. We better have a more general name for the duplicated image. The second is the frame range. The duplication of full stack is better be applicable for stacks with any number of frames, not limited to 15-frames stacks. We can construct the option string of the second argument as shown below to allow the general applicability of the macro.

```
options = "title = dup.tif duplicate range=1-" + nSlices
```

nSlices is a macro function that returns the number of frames or slices in the current stack. This macro function allows the duplication all frames of a stack, regardless of the number of frames within that stack.

We can now replace the second argument for image duplication by this new variable `options`.

```
run("Duplicate...", options);
```

Exercise 2.2

Create a new script tab and write the code below (please ignore the line numbers when you write the code). Run the code with various stacks with different slice or frame numbers and confirm that this short macro successfully duplicate stacks with any slice or frame numbers.

```
print(nSlices);
options = "title=dup.tif duplicate range=1-" + nSlices;
print(options);
run("Duplicate...", options);
```

Below is the upgraded code. All the lines previously commented out were removed, and line 10 was inserted for preparing options for the `Duplicate` command. In addition, we added line 19–24 for closing all images that are not needed anymore.

code/code_block2_recordNucSegV3.ijm

```
1  orgID = getImageID();
2  run("Gaussian Blur...", "sigma=1.50 stack");
3
4  setAutoThreshold("Otsu dark");
5  setOption("BlackBackground", true);
6  run("Convert to Mask", "method=Otsu background=Dark calculate black");
7  run("Analyze Particles...", "size=800-Infinity pixel circularity=0.00-1.00 show=Masks display exclude clear include stack");
8  dilateID = getImageID();
9  run("Invert LUT");
10 options = "title = dup.tif duplicate range=1-" + nSlices;
11 run("Duplicate...", options);
12 erodeID = getImageID();
13 selectImage(dilateID);
14 run("Options...", "iterations=2 count=1 black edm=Overwrite do=Nothing");
15 run("Dilate", "stack");
16 selectImage(erodeID);
17 run("Erode", "stack");
18 imageCalculator("Difference create stack", dilateID, erodeID);
19 selectImage(dilateID);
20 close();
21 selectImage(erodeID);
22 close();
23 selectImage(orgID);
24 close();
```

We now have a block that segments nucleus rim. Save this code, and we are done with the second block of the workflow.

2.4.1.3 Block 3: Intensity Measurement Using Mask

Using the isolated nucleus rim image, we can specify the region for measuring the fluorescence intensity in the Lamin receptor channel. This will be the third block of the workflow.

Before start writing the third block of the workflow, we do a small preparation. We merge the rim-segmented stack and the Lamin receptor stack to create a multi-channel stack, which will be used as the input image data of this third block. Open the rim binary image (if you closed it already, run the second block macro again to regenerate it!) and the Lamin receptor image.

Two stacks can be merged to a two channel image stack by the following command.

`[Image > Color > Merge Channels...]` In the dialogue window, assign red color (C1) to the nucleus channel (nucleus rim binary image), and green color (C2) to the NPT channel. Make sure that "Create composite" is ticked. Clicking "OK" button, you will have an image stack that looks like ◘ Fig. 2.4.

We are now ready to start writing the third block of the workflow. Please follow the steps below using GUI. When you become sure with the operations, record your operations using Command Recorder.

1. `[Image > Color > Split Channels...]`
2. `[Analysis > Set Measurements...]`
   - You will see a dialog window with many check boxes (◘ Fig. 2.5). Among the parameters to be measured, tick at least `Area`, `Mean gray value` and `Integrated density`. Integrated density is the sum of all pixel values.
3. Activate the rim image and do `[Edit > Selection > Create Selection]`
   - This selects the background, not the rim.
4. `[Edit > Selection > make Inverse]`
   - Inverting the selection, now we are selecting the nucleus rim.

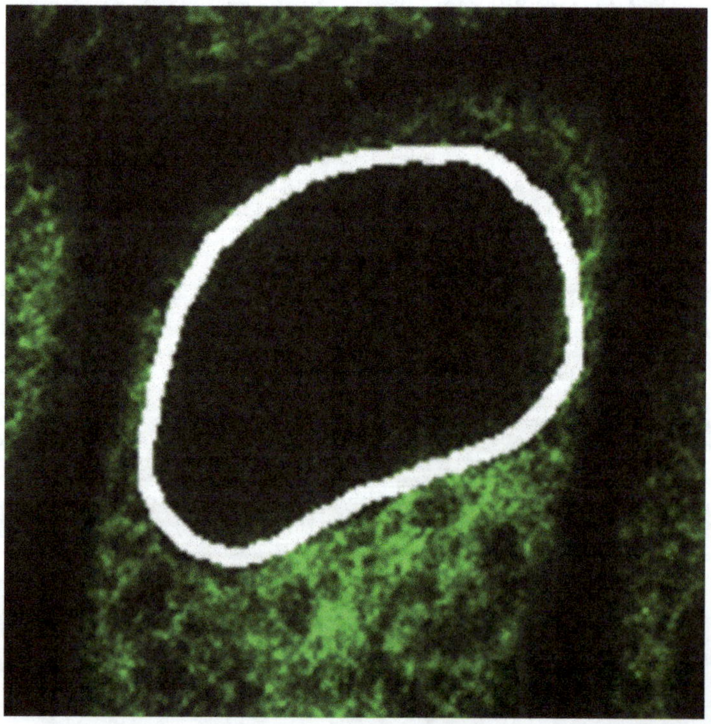

◘ **Fig. 2.4** Merged image of the segmented nucleus rim and the lamin receptor channel

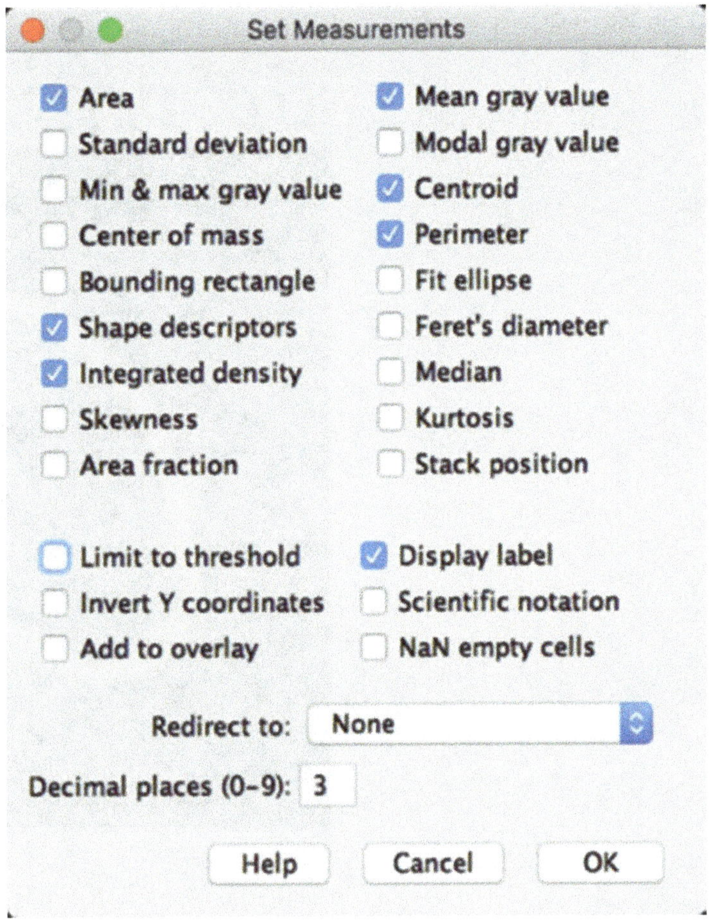

Fig. 2.5 Measurement settings

5. Activate the Lamin receptor image (C2) and then [Edit > Selection > Restore Selection]
6. `[Analyze > Measure]`

Selection of the rim should look like Fig. 2.6.

You will then see results in the Results table such as shown in Fig. 2.7.

When you record these procedures by Command Recorder, the code will look like shown below. Create a new tab in the Script Editor and copy & paste (or it's possible to do the same by clicking "create" button in the Recorder).

```
run("Split Channels");
run("Set Measurements...", "area mean centroid perimeter shape integrated display redirect=None decimal=3");
selectWindow("C1-Composite");
run("Create Selection");
run("Make Inverse");
selectWindow("C2-Composite");
run("Restore Selection");
run("Measure");
```

Fig. 2.6 ROI selection of nucleus rim

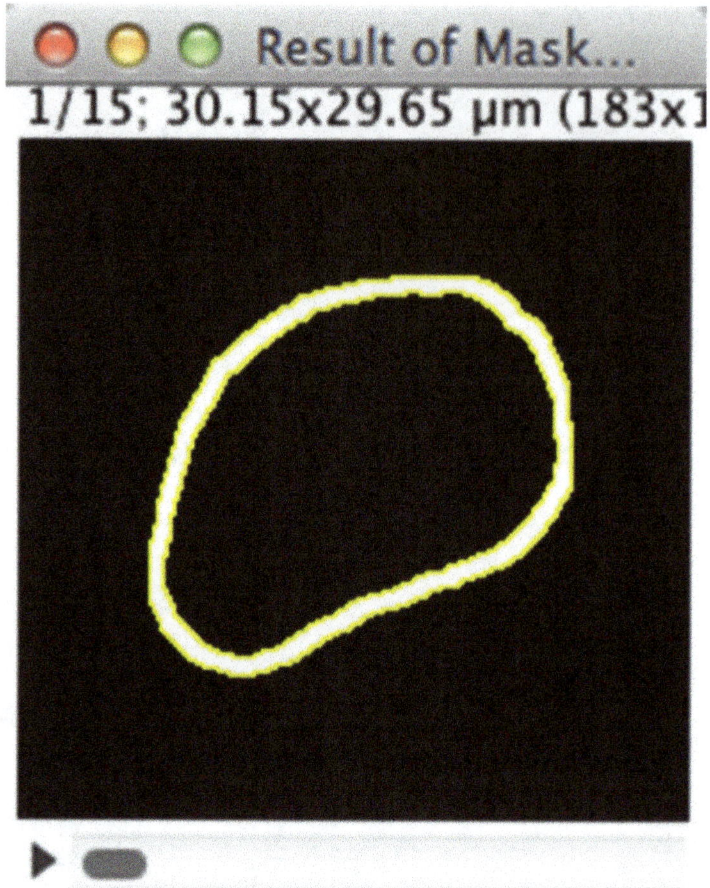

Results

	Label	Area	Mean	X	Y	Perim.	Circ.	IntDen	RawIntDen	AR	Round	Solidity
1	ch1signal.tif:1	46.894	40.511	15.042	14.356	114.752	0.045	1899.721	70003	1.350	0.740	NaN

Fig. 2.7 Results output

In the 1st line, we split the multichannel stack to do processing individually. In the 3rd and the 6th lines, specific window titles are used. Just like we did in the first block, we need to convert these lines by composing window title of stacks for individual channel by adding prefixes. We also need to acquire their image IDs. For composing window titles, we just need to reuse the code we wrote already in the ► Sect. "Block 1: Splitting Channels".

code/code_block3_measurements.ijm

```
orgName = getTitle();
run("Split Channels");
c1name = "C1-" + orgName;
c2name = "C2-" + orgName;
selectWindow(c1name);
c1id = getImageID();
selectWindow(c2name);
c2id = getImageID();
```

```
9  opt = "area mean centroid perimeter shape integrated limit dis-
   play redirect=None decimal=3";
10 run("Set Measurements...", opt);
11 selectImage(c1id);
12 run("Create Selection");
13 run("Make Inverse");
14 selectImage(c2id);
15 run("Restore Selection");
16 run("Measure");
```

- Line 1: We first need to capture the title of the multi-channel image.
- Line 2: Then the channels are separated into two stacks.
- Line 3–4: Since we know the rule of how the resulting image stack names are, we construct titles each for channel 1 and channel 2.
- Line 5–8: We then acquire image IDs.
- Line 9–10: To be more explicit, we compose the measurement options as `opt` in line 9 and then use that variable `opt` as an argument for `Set Measurements` in line 10.
- Line 11: Activate nucleus rim image using ImageID captured in line 6, instead of using `selectWindow`.
- Line 12–13: Create nucleus rim ROI (a selection).
- Line 14: Activate Lamin receptor image using ImageID captured in line 8.
- Line 15: Restore the ROI created in line 13.
- Line 16: We measure the region specified by the ROI created above.

Exercise 2.3

Merge the nucleus rim and the Lamin receptor image stacks as described in the beginning of this section and test the code `code_block3_measurements.ijm` to measure the fluorescence intensity of the nuclear rim.

2.4.2 Integration: The Measurement Over Time

The code above measures only one time point. To measure the intensity changes over time, we need to add **looping** from line 11 to 16 in `code_block3_measurements.ijm` to repeat the measurement over time frames. For this, we need to modify the code by adding a **for-loop**.

code/code_block3_MeasurementOverTime.ijm

```
orgName = getTitle();
run("Split Channels");
c1name = "C1-" + orgName;
c2name = "C2-" + orgName;
selectWindow(c1name);
c1id = getImageID();
selectWindow(c2name);
c2id = getImageID();
```

```
9  opt = "area mean centroid perimeter shape integrated limit display
   redirect=None decimal=3";
10 run("Set Measurements...", opt);
11 for (i =0; i < nSlices; i++){
12     selectImage(c1id);
13     setSlice(i + 1);
14     run("Create Selection");
15     run("Make Inverse");
16     selectImage(c2id);
17     setSlice(i + 1);
18     run("Restore Selection");
19     run("Measure");
20 }
```

In this updated code, following 4 lines were added for looping through the time lapse frames and measure successively.

- A new line was inserted at line 11 to define the condition of for-looping.
- A new line was inserted at line 13 to activate a specific frame in the nucleus rim stack.
- A new line was inserted at line 17 to activate a specific frame in the stack.
- A curly brace was added at line 20 to close the looping.

Exercise 2.4

Merge rim and Lamin receptor image stacks and test the code `code_block3_MeasurementOverTime.ijm` to see if it measures the intensity of nucleus rim over time frames.

If you see 15 lines of measurement values in the Results window, you are successful.

2.4.3 Integrating Segmentation and Measurements

Finally, we can assemble three blocks of code: the channel splitting block, the segmentation block and the intensity measurement block. As the third block, the intensity measurement block, starts with a two-channel stack (nucleus rim segmentation image and the Lamin receptor signal image), all we need to do is to insert the segmentation block between line 4 and line 5 of block 3 code `code_block3_MeasurementOverTime.ijm`.

Instead of copy and pasting the segmentation block to the measurement block, a better way to do this is to convert the segmentation block to a user-defined function. Like all the macro commands that you see in the Build-in ImageJ macro function reference, we can create our own function by ourselves. We briefly learn how to write a custom function with a simple example.

If we have a code like below:

```
a = 10;
b = 20;
c = a + b + a * b ;
print(c);
```

Evidently, “230” will be printed in the log window. Now, We can convert this formula to a custom function `calc1` that does the calculation in line 3.


```
1  a = 10;
2  b = 20;
3  c = calc1(a, b);
4  print(c);
5
6  function calc1(n, m){
7    return n + m + n * m;
8  }
```

Three lines were added to the original code. Line 6 declares a new user-defined function named “calc1”. It takes two arguments, `n` and `m`. Commands between curly braces is the content of this function, and in this case there is only one line that returns a value. To be more explanatory, this function can be rewritten as follows to do the same thing.

```
function calc1(n, m){
    answer = n + m + n * m;
  return answer;
}
```

Exercise 2.5

1. Modify the code above so that the function `calc1` calculates m to the power of n. Use the build-in command pow(m, n).
2. Change the name of function to `calc2` and run the code. If there is error, fix the code.

In a similar way, we can convert the segmentation block to a single custom function that takes an ImageID as input, does pre-processing, does segmentation, and then returns an ImageID of the segmented image as the output. Here is the code:

code/code_block2_recordNucSegV3_function.ijm

```
function nucseg(orgID){
  //orgID = getImageID();
  selectImage(orgId);
  run("Gaussian Blur...", "sigma=1.50 stack");

  setAutoThreshold("Otsu dark");
  setOption("BlackBackground", true);
  run("Convert to Mask", "method=Otsu background=Dark calculate
  black");
  run("Analyze Particles...", "size=800-Infinity pixel circular-
  ity=0.00-1.00 show=Masks display exclude clear include stack");
  dilateID = getImageID();
  run("Invert LUT");
  options = "title = dup.tif duplicate range=1-" + nSlices;
```

```
13    run("Duplicate...", options);
14    erodeID = getImageID();
15    selectImage(dilateID);
16    run("Options...", "iterations=2 count=1 black edm=Overwrite
      do=Nothing");
17    run("Dilate", "stack");
18    selectImage(erodeID);
19    run("Erode", "stack");
20    imageCalculator("Difference create stack", dilateID, erodeID);
21    resultID = getImageID();
22    selectImage(dilateID);
23    close();
24    selectImage(erodeID);
25    close();
26    selectImage(orgID);
27    close();
28    run("Clear Results");
29    return resultID;
30  }
```

Only several lines were added to the original `code_block2_recordNucSegV3.ijm`.

- In line 1, we declare that this is a custom function named `nucseg` that takes a single argument `orgID`. In the original code, `orgID`, which is the imageID of the histone channel image, was captured using `getImageID` command.
- Line 2 is commented out. This is because We do not need to do `getImageID` since the imageID of the histone channel image is provided through the argument of the function.
- Line 3 is inserted, to explicitly activate the image with id `orgId`.
- One line is inserted at line 21, to capture the imageID of the resulting image stack—the mask of nuclear rim—of Image Calculation in line 20. This imageID is named as a variable "resultID" in the function and is returned in the line 29 as the final output of the function.
- `run("Clear Results");` is added at the bottom (line 28) to clear the results table, as we want to have only the results of intensity measurement later.
- In the last line, a curly brace is added to mark the boundary of function.

We can paste this function `nucseg(orgID)` below the intensity measurement macro, and call this function to segment the nucleus rim. In below, I show only the part in the block 3 intensity measurement where function call was added. line 7 to line 8 was inserted to `code_block3_MeasurementOverTime.ijm`.

```
1  orgName = getTitle();
2  run("Split Channels");
3  c1name = "C1-" + orgName;
4  c2name = "C2-" + orgName;
5
6  selectWindow(c1name);
```

```
7  nucorgID = getImageID();
8  nucrimID = nucseg(nucorgID);
9
10 selectWindow(c2name);
11 c2id = getImageID();
12 opt = "area mean centroid perimeter shape integrated display
   redirect=None decimal=3";
13 run("Set Measurements...", opt);
14 for (i =0; i < nSlices; i++){
15     selectImage(nucrimID);
16     setSlice(i + 1);
17     run("Create Selection");
18     run("Make Inverse");
19     selectImage(c2id);
20     setSlice(i + 1);
21     run("Restore Selection");
22     run("Measure");
23 }
```

In line 6 and 7, the image ID of the nucleus (histone) channel is captured. As we do not know if the nucleus channel image stack is the top window, we explicitly call it to the top by `selectWindow`, and then get its ImageID. This ImageID `nucorgID` is then passed to the segmentation function in line 8 (`nucseg( nucorgID)`).

After the image segmentation is done in the function `nucseg`, the ImageID of segmentation result is returned. We capture this ImageID by a variable `nucrimID`. From there, everything is same like we already coded, except that the image selection at line 15 now uses `nucrimID`.

Here is the final code.

code/code_final.ijm

```
1  orgName = getTitle();
2  run("Split Channels");
3  c1name = "C1-" + orgName;
4  c2name = "C2-" + orgName;
5
6  selectWindow(c1name);
7  nucorgID = getImageID();
8  nucrimID = nucseg(nucorgID);
9
10 selectWindow(c2name);
11 c2id = getImageID();
12 opt = "area mean centroid perimeter shape integrated display
   redirect=None decimal=3";
13 run("Set Measurements...", opt);
14 for (i =0; i < nSlices; i++){
15     selectImage(nucrimID);
16     setSlice(i + 1);
17     run("Create Selection");
18     run("Make Inverse");
19     selectImage(c2id);
```

```
        setSlice(i + 1);
        run("Restore Selection");
        run("Measure");
    }

function nucseg(orgID){
        selectImage(orgId);
        run("Gaussian Blur...", "sigma=1.50 stack");

        setAutoThreshold("Otsu dark");
        setOption("BlackBackground", true);
        run("Convert to Mask", "method=Otsu background=Dark calculate black");
        run("Analyze Particles...", "size=800-Infinity pixel circularity=0.00-1.00 show=Masks display exclude clear include stack");
        dilateID = getImageID();
        run("Invert LUT");
        options = "title = dup.tif duplicate range=1-" + nSlices;
        run("Duplicate...", options);
        erodeID = getImageID();
        selectImage(dilateID);
        run("Options...", "iterations=2 count=1 black edm=Overwrite do=Nothing");
        run("Dilate", "stack");
        selectImage(erodeID);
        run("Erode", "stack");
        imageCalculator("Difference create stack", dilateID, erodeID);
        resultID = getImageID();
        selectImage(dilateID);
        close();
        selectImage(erodeID);
        close();
        selectImage(orgID);
        close();
        run("Clear Results");
        return resultID;
    }
```

2.5 Results and Conclusion

The final output is a list of nucleus rim intensity values for each time point in Results window. These values can be saved in a CSV file and further analyzed using other software tools more suited for data analysis such as R or Python. Here, to summarize the analysis in this chapter, we plot the changes in the total fluorescence intensity over time using ImageJ Macro `code_plotResults.ijm` (Fig. 2.8). The code appears after the paragraph below.

The plot in Fig. 2.8 shows an increase in total fluorescence intensity by 1.3-folds in the initial five time points, and then it becomes mostly constant. To know the baseline level intensity more precisely, it might be better to start the imaging and measurement from an earlier time point. In addition, ideally, more measurements could be done with other nuclei to compute an averaged curve for a more reliable results.

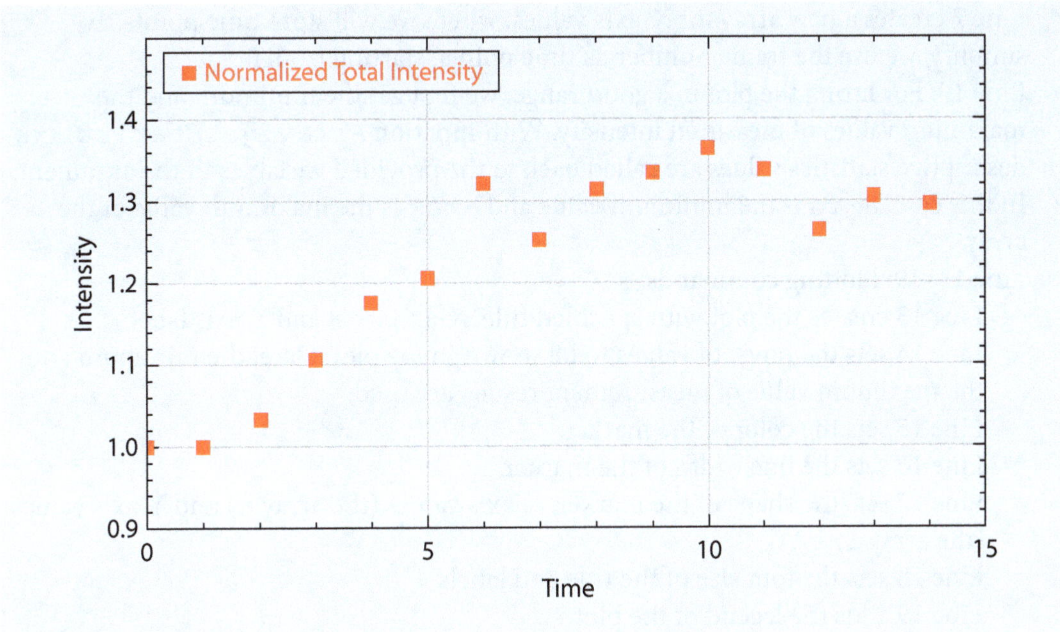

Fig. 2.8 Changes in the total fluorescence intensity over time

Here is the code for this plotting. Explanations follow.

code/code_plotResults.ijm

```
1  // store normalized total intensity values in an array
2  intA = newArray(nResults);
3  for (i = 0; i < nResults; i++)
4    intA[i] = getResult("RawIntDen", i) / getResult("IntDen", 0);
5
6  //prepare x-axis values
7  t = Array.getSequence(intA.length);
8
9  // get the statistics of the total intensity array.
10 Array.getStatistics(intA, amin, amax, amean, astdDev);
11
12  // Create the plot
13 Plot.create("Total Intensity at Nuclear Membrane", "Time", "Inten-
   sity");
14 Plot.setLimits(0, intA.length, amin * 0.9, amax * 1.1);
15 Plot.setColor("red", "red");
16 Plot.setLineWidth(3);
17 Plot.add("circle", t, intA);
18 Plot.setFontSize(14);
19 Plot.addLegend("Normalized Total Intensity");
```

- Line 2 creates a new array for storing intensity measurements listed in the Results table. `nResults` is a build-in function that returns number of rows in the table. This array will be the Y-axis value in the plot.
- Line 3–4 gets the result of non-calibrated integrated density (Column "RawIntDen") of each row, and divide that value by the integrated density at the time point 0 (the first frame).

- Line 7 creates a new array for X-axis values, where we will store time points. To simplify, we use the frame number as time points, starting from 0.
- Line 10: For fitting the plot in a good range, we first get the minimum and the maximum values of measured intensity. With function `Array.getStatistics`, descriptive statistics values are called back to the provided variables in the argument. In this case, `amin` is the minimum value and `amax` is the maximum value of the array.
- Line 13–19: Plotting commands.
  - Line 13 creates the plot with specified title, X-axis label and Y-axis label.
  - Line 14 sets the range of values to be shown in the plot. Here, the minimum and the maximum value of measurement results are used.
  - Line 15 sets the color of the marker.
  - Line 16 sets the line width of the marker.
  - Line 17 sets the shape of the marker, X-axis values (the array `t`) and Y-axis values (the array `intA`)
  - Line 18 sets the font size of the title and labels
  - Line 19 adds the legend of the plot.

Take Home Message

To measure the changes in the fluorescence intensity over time at the nuclear membrane, we post-processed segmented image of nucleus by mathematical morphology processing "Erosion" and "Dilation" to create a mask for the region-of-interest. In the same way, boundaries of biological structures can be segmented and analyzed.

2.6 Exercise Answers

2.6.1 Exercises 2.1–2.4

In these exercises, one only needs to follow the instructions.

2.6.2 Exercise 2.5

1. Modify the code above so that the function `calc1` calculates m to the power of n. Use the build-in command pow(m, n).

```
a = 10;
b = 20;
c = calc1(a, b);
print(c);

function calc1(n, m){
  return pow(m , n);
}
```

2. Change the name of function to `calc2` and run the code. If there is error, fix the code.
   Answer: Be sure that`calc1` in line 3 needs to be replaced by `calc2` as well.

```
1 a = 10;
2 b = 20;
3 c = calc2(a, b);
4 print(c);
5
6 function calc2(n, m){
7   return pow(m, n);
8 }
```

Acknowledgements This workflow was initially developed together with Andreas Boni for teaching in a practical course at EMBL Heidelberg. We are grateful to his contributions. The same topic was taught in many courses during last five years and we thank all the feed-backs we received from those teaching sessions from participants. We thank Andreas Boni, Nathalie Daigle, and Jan Ellenberg for providing the sample image data. We thank Christian Tischer (EMBL Heidelberg) for reviewing this chapter.

Bibliography

Boni A, Politi AZ, Strnad P, Xiang W, Hossain MJ, Ellenberg J (2015) Live imaging and modeling of inner nuclear membrane targeting reveals its molecular requirements in mammalian cells. J Cell Biol 209(5):705–720. ISSN: 0021-9525. https://doi.org/10.1083/jcb.201409133. http://www.jcb.org/lookup/doi/10.1083/jcb.201409133

3D Quantitative Colocalisation Analysis

Bioimage Analysis Series

Fabrice P. Cordelières and Chong Zhang

K. Miura, N. Sladoje (eds.), *Bioimage Data Analysis Workflows*, Learning Materials in Biosciences,
https://doi.org/10.1007/978-3-030-22386-1_3

What You Learn from This Chapter

In this module we will first build a 3D object based colocalisation macro step by step. Then we will practice to adapt and extend the current macro such that it can also work with intensity-based colocalisation methods.

3.1 Introduction

3.1.1 What Is Colocalisation?

Subcellular structures interact in numerous ways, which depend on spatial proximity or spatial correlations between the interacting structures. Colocalisation analysis aims at finding such correlations, providing hints of potential interactions. If the structures only have simple spatial overlap with one another, it is called *co-occurrence*; If they not only overlap but also co-distribute in proportion, it is then *correlation*.

Two proteins are said to be colocalised when the locations of their associated signals (fluorescence) are indistinguishable by the imaging system used, i.e. the distance between signal is below the resolution of the imaging system. For example, in ◻ Fig. 3.1, the seemingly one pair of colocalised objects in low resolution images are actually several pairs of smaller objects in close proximity. Therefore, the diagnosis placed for colocalisation should always be stated relative to a particular resolution and sampling rate. In other words, conclusions that could be drawn from a colocalisation study are:

- **In cell biology**: the two proteins are at the same location;
- **In statistics**: considering the current resolution, it might not be excluded that the two proteins are indeed at the same location.

3.1.2 Which Colocalisation Methods Are There?

In general, when we have a specific application for colocalisation analysis, a few questions should be asked first, and depending on the answers to them, one or more methods should be applied. Here in ◻ Fig. 3.2 is our recipe for which method(s) are appropriate. Specifically, colocalisation may be evaluated visually, quantitatively, and statistically:

- It may be identified by superimposing two images and inspecting the appearance of the combined color. For example, colocalisation of red and green structures can appear yellow. However, this intuitive method can work only when the intensity levels of the two images are similar (see a detailed example in Dunn et al. (2011)). Scatter

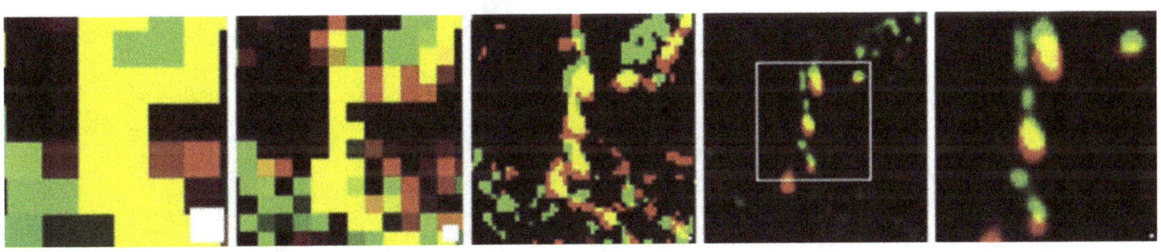

◻ **Fig. 3.1** (*left to right*) The same objects imaged with varying resolutions, expressed as a fraction of the highest resolution: 1/16, 1/8, 1/2, 1, and a zoom-in view (*white frame*). The white pixel in the bottom right corner depicts the sampling rate (pixel size) adapted to each resolution. The seemingly one pair of colocalised objects are actually several pairs, as shown in images with higher resolution

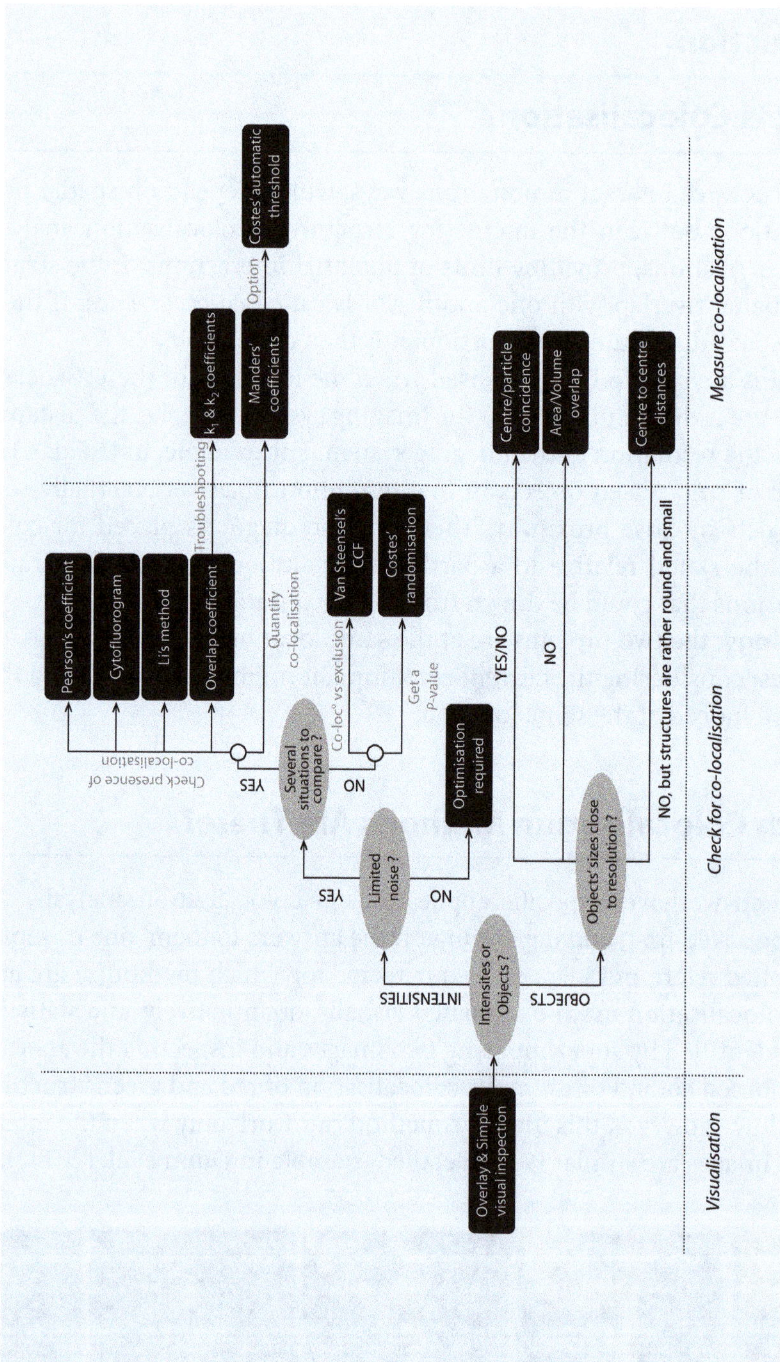

Fig. 3.2 Our guideline decision tree

plot of pixel intensities from the two images also qualitatively indicates colocalisation, e.g. the points form a straight line if the two structures correlate. But visual evaluation does not tell the degree of colocalisation, nor if it is true colocalisation at all.

- In general, two categories of quantitative approaches to colocalisation analysis can be found: intensity based correlation methods and object based methods. Intensity based methods compute global measures about colocalisation, using the correlation information of intensities of two channels. Several review papers have been published during the last decade, where coefficients' meaning, interpretation, guide of use, and examples for colocalisation are given (Bolte and Cordelières 2006; Cordelières and Bolte 2014; Dunn et al. 2011; Zinchuk and Zinchuk 2008). Tools for quantifying these measures can be found in many image analysis open-source and commercial software packages, to name just a few: Fiji's JACoP plugin and Coloc 2, CellProfiler, BioImageXD, Huygens, Imaris, Metamorph, Volocity. Most object-based colocalisation methods first segment and identify objects, and then account for objects' inter-distances to analyze possible colocalisation. Usually, two objects are considered colocalised, if the centroids of the objects are within certain distance (Bolte and Cordelières 2006; Cordelières and Bolte 2014; Obara et al. 2013), or if two objects with certain percentage of area/volume overlap (Rizk et al. 2014; Wörz et al. 2010). We will implement some specific methods for both categories in two case studies described in ▶ Sects. 3.4 and 3.5.
- Colocalisation studies generally should perform some statistical analysis, in order to interpret whether the found co-occurrence or correlation is just a random coincidence or a true colocalisation. A common method is Monte-Carlo simulations Fletcher et al. (2010) but it is computationally expensive. Recently a new analytical statistics method based on Ripley's K function is proposed and included as an Icy plugin, Colocalisation Studio (Lagache et al. 2013).

3.1.3 Some Image Preprocessing Tips You Should Keep in Mind

Talking about colocalisation, we often also think about deconvolution. Careful image restoration by deconvolution removes noise and increases contrast in images, improving the quality of colocalisation analysis results. Noisy images may generate unwanted "matching pixels": it should therefore be handled with great care. High contrast is always a plus, especially when trying to delineate structures for individual object's colocalisation determination. In Fiji, you can find several plugins for these tasks to try on your images, such as:

- Parallel Iterative Deconvolution (fiji.sc/Parallel_Iterative_Deconvolution), where the point spread function (PSF) can be estimated using the Diffraction PSF 3D plugin (fiji.sc/Diffraction_PSF_3D). An example can be found in ◘ Fig. 3.3.
- DeconvolutionLab and DeconvolutionLab2 (Sage et al. 2017).
- To further remove background noise, you can try `[Process -> Subtract Background]` (for our images, the rolling ball radius can be set to 10 pixels)

However, deconvolution is not the focus of this module. Therefore, we would assume that the images to be processed during this module are either already deconvolved or are acquired with high image quality without the need of deconvolution. Other issues that may need to be dealt with during the preprocessing include: illumination correction, noise removal, background or artifacts disturbance. Since we have already been practicing techniques to handle these situations, here we would as well rather not to discuss them.

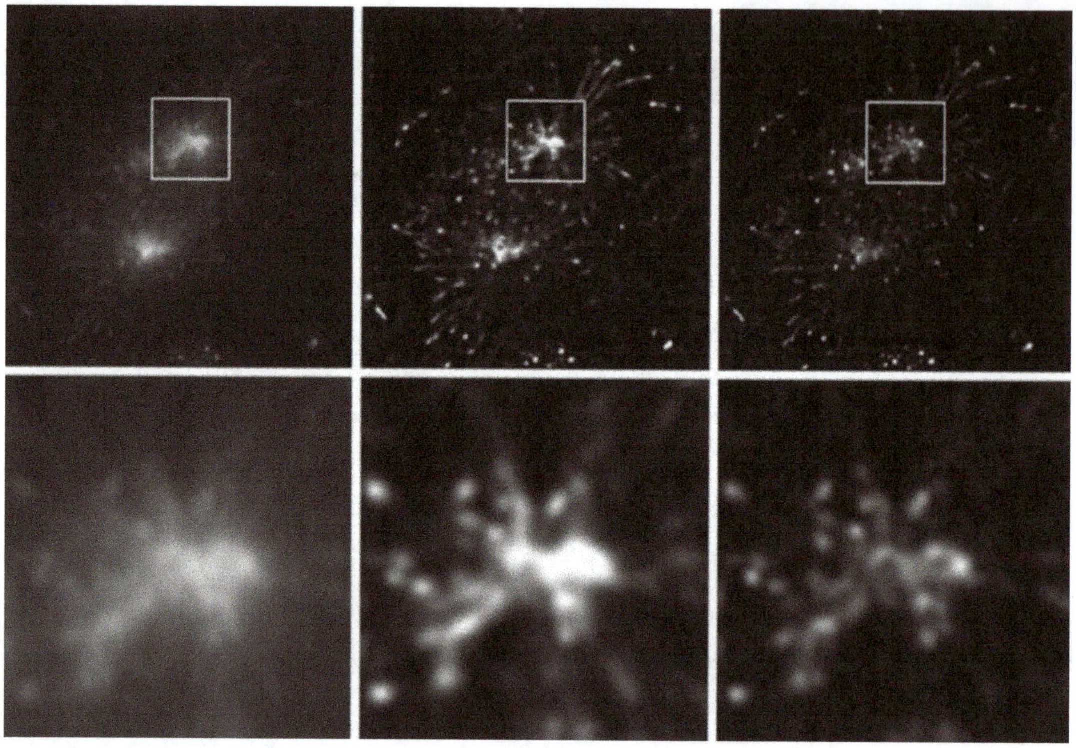

Fig. 3.3 Example images before (*left*) and after (*middle/right*) two deconvolution algorithms (Richardson & Lucy, 200 iterations, middle; Meinel, 10 iterations, right). Lower row presents a magnification of upper figures, centered on the upper mitotic spindle pole

It is worth noting briefly here that before this step, several points should be taken care of during the image acquisition and collection part.

- To have the imaging hardware set up appropriately. That is, to adjust the exposure time, detector gain and offset so as to be able to detect the dimmest structures without saturating the brightest structures.
- To check for chromatic aberrations one uses small beads that are fluorescent in many colors and thus should 100% co-localize with themselves. If they appear shifted you have to realign your microscope or account for the shift during the analysis.
- To appropriately control bleed-through.

3.2 Datasets

Let's first have a look at the data. The images display a large number of dash-like structures. The life scientist informs us they result from the labeling of two proteins which both locate at the most dynamic end of microtubules: they belong to the + TiPs family (plus-end tracking proteins). When overlaying both images (EB1 on channel 1 and CLIP-170 on channel 2, available in this module's folder: Images/+ TIPs.zip and as a smaller version Images/+ TIPs_mini.zip), it seems that the overlap is only partial, which justifies the scientist's first impression. Now we have to "put numbers" and try to evaluate this partial colocalisation to either demonstrate it or prove it to be a simple visual artifact (Fig. 3.4).

3

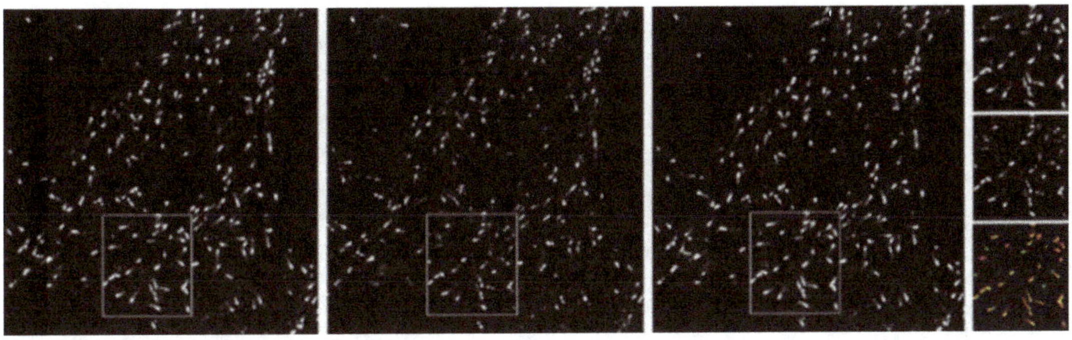

Fig. 3.4 Dataset on which colocalisation will be evaluated (Left: EB1, Middle: CLIP-170, Right: Overlay, Right-most: magnified sub-images)

Fig. 3.5 Don't trust your eyes and brain on colocalisation!

Of course, most of us would rule out this latter hypothesis (Fig. 3.5). So let's take an example from Fiji's website: do you think some of the red and green pixels colocalise on this image?

You may try zooming at the image and realize that yellow pixels are in fact resulting from the close proximity between green and red pixels: our brain simply blends one tone into the other. In case you do not believe this is an optical trick, try opening it under ImageJ then moving the cursor over the "yellowish" pixels: the status bar will display the values of the green and red components of the image. As a result, you'll figure out that red pixels are not green and the reverse way round.

Word of advise: in colocalisation studies, as you can't trust your eyes and the brain that lays (or lies) behind, better be ready to build a proper quantification strategy!

In addition to the requirement of a proper quantification strategy, this example also points out the need for a well characterized dataset. When building a workflow,

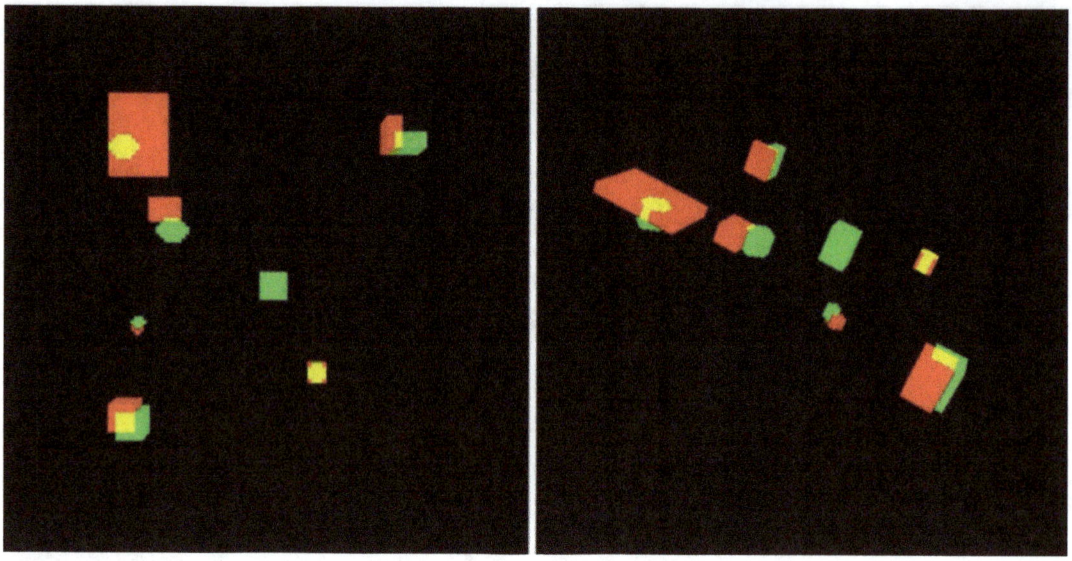

Fig. 3.6 Synthetic 3D dataset from two views

the image analyst might benefit from use of a synthetic dataset. This somewhat barbarian terms designate a computer-generated dataset where all parameters are human controlled. In microscopy, the images result from single captions of a scene, which are impaired by the optics and corrupted by noise. To simplify the prototyping process, we could generate two images, one per channel, containing simple shapes (circles, rectangles etc...) which sizes, and degree of overlap are controlled. Such a dataset, extended to the 3D case is provided. Figure 3.6 shows two 3D views of the synthetic dataset with two channels, where channel 1 (*red*) has six objects and channel 2 (*blue*) seven. Each object in channel 1 has different level of spatial overlap with one of the objects in channel 2. The synthetic dataset can be found in this module's folder (Images/Synthetic.zip).

3.3 Tools

- Fiji
  - Download URL: ▶ https://imagej.net/Fiji/Downloads

3.4 Workflow 1: Objects Overlap Volume Quantification

Let's imagine a typical conversation between an image analyst and a life scientist who has a colocalisation analysis request for the dataset to be reviewed under ▶ Sect. 3.2:

> I've got a set of two-channel 3D images where objects are overlapping. I think the overlap might not be the same from object to object.
>
> Therefore, I would like to quantify the overlap and get a map of quantification.
>
> *A user comes to the Facility, asking for help*

3.4.1 Step 0: Building a Strategy

So how would we tackle the problem?

The key words here are: objects, overlapping, map of quantification. So the first task is to identify objects of interest in the image, i.e. segmentation. Then, overlapping regions between objects in the two channels should be extracted. In order to obtain a map of quantification on overlap, one of the quantification metrics could be calculating the volume of objects and overlapping regions. ◘ Figure 3.7 and ◘ Table 3.1 summarizes the steps that need to be implemented in the macro.

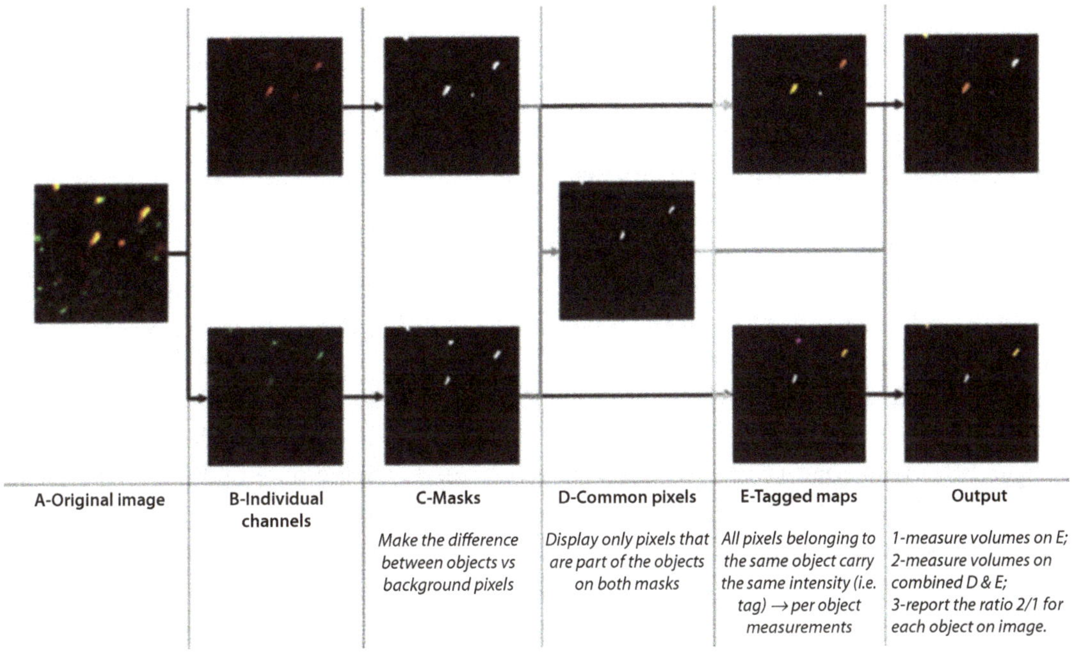

◘ **Fig. 3.7** A schematic drawing of the full workflow: the first task is to identify objects of interest in the image, i.e. segmentation. Then, overlapping regions between objects in the two channels should be extracted. In order to obtain a map of quantification on overlap, one of the quantification metrics could be calculating the volume of objects and overlapping regions

◘ **Table 3.1** What needs to be done?

Step	What needs to be done?	What for?
1	Normalize the data name	Split & rename channels
2	Tag the objects	Segment & label 3D objects
3	Isolate the overlapping parts	Segment overlapping regions
4	Retrieve volumes	Calculate & store object volumes
5	Generate outputs	Display & visualize volume ratio
6	Make the macro user friendly	Create user interface

Since most of the operations will be applied the same way on both channels, or on different objects/regions, the most efficient scripting strategy would be to define `functions` that consist of several macro commands such that a specific operation is performed. And these functions could be simply called by the function name plus arguments as many times as needed in the macro, instead of repeating the macro command lines. This way improves the readability of the macro thus the re-usability as well.

3.4.2 Step 1: Normalize the Image Names

Aim:

- The channels should be split so that images are processed independently.
- The first channel should be named as "Channel1", the second "Channel2".
- Generate a user-defined function to perform the task: `''normaliseNames''` with proper arguments, if needed.

When automating an image processing workflow, a major challenge is to make the macro re-usable and independent about the input images' naming convention. We will be facing an additional complication as the image we are working on is a composite, made of several channels. Processing will be applied to each single channel independently, once they will have been splitted. A proper strategy should be designed to keep track of the original image's title (ie, its name) and the subsequent channels' image obtained after separating them. The following section will handle this step in a simple way by renaming the original and the subsequent images with pre-defined names. This step could be seen as "normalizing the images' names" for better handling.

We will use the function [`Plugins > Macros > Record`] to track the sequential operations that have been applied; clean up and convert the recorded macro commands to our own defined function in the macro file. Some of the functions, for example to retrieve an image's title, are not recordable. In such case, making a simple word-based search on ImageJ's Built-in Macro Functions page might help. `getTitle( )` function returns the title of the active image as a String that can be stored into a variable. Let's call it `''ori''` as it is used to store the name of the "original" image.

The original image is a composite. For the kind of processing we are planning to do, each channel should be handled separately. Channels are splitted by using the function [`Image > Color > Split channels`] that is macro recordable. ImageJ/Fiji applies specific naming rules to each image when splitting the channels: the resulting titles are built by concatenating the letter "C" (for channel), the number of the channel (starting at 1), a dash, and finally the original image's title. We therefore have a way to precisely activate the first or the second channel. However, to make the process a bit easier, we rename each channel in the form of "ChannelXX".

Finally, once all operations have been recorded, and the code cleaned up, the corresponding lines can be encapsulated into a function. To generate a function, you simply need to append the keyword "`function`" with a proper name, a pair of empty parentheses (or contain required parameters) and a pair of curly brackets: the few lines of code we've just written should be copied/pasted in between.

To summarize, here is your working plan to implement the code:

Working Plan:

1. Split the image into its channels
2. Select the first channel
3. Rename the image "Channel1"
4. Select the second channel
5. Rename the image "Channel2"
6. Pack everything into a function, thinking about the proper arguments, if any, that should be entered for the process to be run

The correct code of the function is:

```
//Split the channels and rename them in a normalised way
function normaliseNames(){
ori= getTitle ();
run ("Split Channels");
selectWindow ("C1-"+ori);
rename ("Channel1");
selectWindow ("C2-"+ori);
rename ("Channel2");
}
```

Now, in order to run the macro properly with this function, we still need a line in the main code body of the macro to call it. For this particular step, it is straightforward to call the function, as follows:

▪▪ The main macro till Step 1

```
//----------------------------------------
// Main macro part

//Step 1
normaliseNames();

//----------------------------------------
}
```

Something that is as important as writing the macro itself is to comment in detail each step you do. This way, it not only helps to remind what the macro does, but also helps to reuse and adapt existing macros. The latter is much better practice than always writing things from scratch whenever you need to do some analysis. You will see later why it is so.

3.4.3 Step 2: Tag the Objects

Aim:

- For each channel, have a tagged map: all pixels from an object should carry a same tag/intensity.
- The tagging should be made in 3D.
- Generate a user-defined function to perform the task: ``isolateObjects'' with proper arguments, if needed.

Hints:

- We need to differentiate background pixels from objects' pixels: thresholding is one way to achieve this.
- ImageJ built-in functions are not the only ones being recordable.

Well, the aim being explicit, one can already see this part of the workflow is not specific to a single channels: both channels will go through this process. This is therefore a good candidate for building a function, that will be called twice to process each of them, and name as "isolateObjects". The remaining question is how the process would differ from one image to another?

First, we have to define what an object is. Based on the image, we could define an object as a group of contiguous pixels, for which each pixel has an intensity above a certain threshold value.

The latter step is easy to achieve, as long as the background (i.e. the non-object voxels) is distributed within a well defined range of intensities. The function to be used lies in the [Image > Adjust > Threshold...] menu. This function is macro recordable: the recorder should contain the run(``Threshold..'') instruction. Note, however, that this function is commented. Un-commenting and running it won't do much: this function is only aimed at displaying the dialog box, not to perform any processing. However, it is useful, for instance, to pop-up the window before asking a user to set the threshold values. Once more, we will have to refer to ImageJ's Built-in Macro Functions page and look at threshold-related instructions. For any kind of process, two types of functions might be found: setters, to apply parameters and getters to retrieve them. In the specific case of thresholding, the two types exist, in the form of setThreshold(lower, upper) and getThreshold(lower, upper). This information will show its use later on in this module.

The first step, grouping adjacent voxels into objects might seem to be a bit tricky. This is however the kind of processing that goes behind the scenes when using the [Process >Analyze Particles...] function. But how does this work ? First, a destination image is created: we name it as"tagged map" (see Fig. 3.8). Each thresholded pixel from the original image is screened in turn, from the top-left corner to the bottom-right corner, line by line. The first thresholded pixel is attributed a tag, i.e. the corresponding pixel

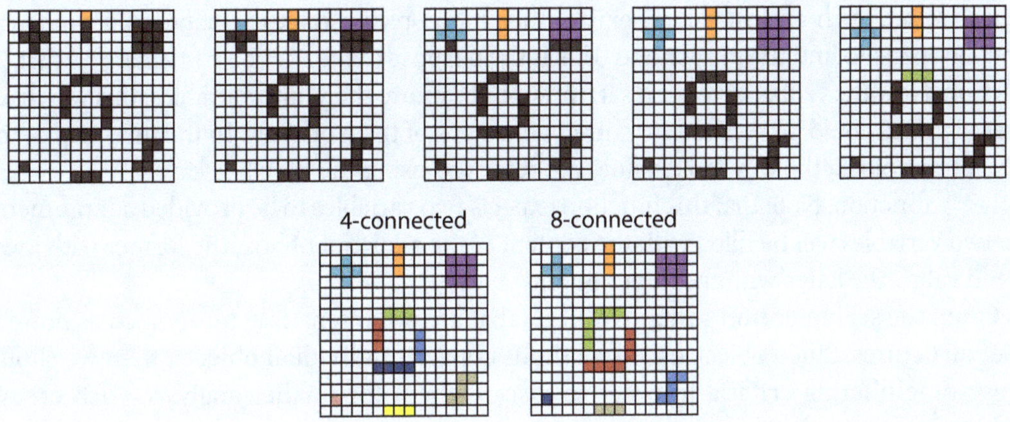

Fig. 3.8 A schematic drawing of converting the segmented object mask into a "tagged map" of these objects, i.e. all pixels that are connected are labeled as the same color or pixel intensity; and those not connected are distinguished by a different color or pixel intensity. There are different configurations of "connected", e.g. 4 or 8 connected in 2D

on the "tagged map" will be given a value of "1". Next pixel (to the right) is examined: if thresholded, a tag will be given. As its direct, left neighbor, as already been given a tag of "1", it receives the same tag: both are part of the same object. This process goes on for the entire image. In case a new thresholded pixel is found, that doesn't have a tagged neighbor, a new tag is created: this pixel is a seed for a new object. Please note that several rules might be applied to define pixels' vicinity: a pixel can be connected only through its sides (4-connected in 2D) or one might also consider its diagonal corners as contact points (8-connected in 2D). Depending on which rule is used, the number of objects might not be the same.

Although being convenient as an ImageJ built-in tool, the "Analyze Particles" function only works in 2D: an object spanning over several slices might end-up being counted several times. In 3D, another tool should be used: 3D-Object counter (aka 3D-OC), which works with the 26-connected voxels rule. As it is fully macro-recordable, its usage from within a macro is straightforward. If using Fiji, the plugin is already installed, if not, you'll have to download it from this link. Note that when using it from a macro: the Fiji team has changed the name of its menu entries as compared to the original author's version: a macro recorded under Fiji might need few changes to run under ImageJ and vice-versa.

We are now familiar with all the required components for isolating objects. Let's do manually all the steps of the workflow, keeping an eye on the macro-recorder. We first need to select the proper window, then launch the threshold adapter and feed the lower threshold to the 3D-OC plugin, and finally ask it to display the objects map (i.e. tagged map).[1]

The implementation is straightforward: as user input is required, the threshold box is first activated. To preset a threshold, assuming fluorescence images, one could press the `Auto` button: the following instruction will be recorded: (`setAutoThreshold ("Default dark");`). Now, we must find a way to display a message box inviting the user to

1 In case you see numbers overlaid onto the tagged map, go to the "`Set 3D Measurements`" menu and uncheck the "`Show numbers`" option.

finely tune the threshold. In the meantime, the macro's execution should be idled until proper values are set. Such an instruction exists, once more, please refer to ImageJ's Built-in Macro Functions page to find the proper one. This is the purpose of the `waitForUser("text")` function. Now the values have been set, we should capture them into variables: this way, user-entered values could be recalled for subsequent steps of the workflow. In the final step of this instructions block, the threshold values are retrieved using the `getThreshold(lower, upper)` function. Note that this function expects two variables to be provided as arguments. The two variables will be filled with the content of the adjustment box, the former with lower bound value, the latter with the upper bound value.

Often the segmentation contains objects that are not interesting for us such as noise or other structures. Since object-based methods concern individual objects, then we should apply some filtering criteria in order to discard them for further analysis. Such criteria could be, for example:

1. (3D/2D/1D) size range of the object-of-interest (in each channel)
2. object shape, e.g. circularity,[2] compactness[3]
3. object location

It should be mentioned that this step greatly influences the colocalisation measurements. We will discuss only size related filtering here. `3D Objects Counter` is able to do this. Then let's select one channel, e.g. Channel 1, to record all the needed operations to convert into the second function, `isolateObjects`. When running [`Analyze > 3D Object Counter`], in the pop-up window there are three parameters of our interest: `Min` and `Max` in the `Size filter` field, and the `threshold`. We could pass along the variable that stores the user specified threshold value in Step 1, i.e. `lower`. Let's suppose that the object of interest should have a size of minimum 5 voxels and maximum 100,000 voxels. This filtering step removes the smallest object from the image. Although they may seem to overlap with objects in the other channel, they are likely to be e.g. noise and their spatial co-occurrence could be coming from randomly distributed particles/noises that are close to each other by chance. Before applying the `3D Objects Counter`, we should check the measurements setting in the [`Analyze -> 3D OC Options`] options window. This is similar to [`Analyze -> Set Measurements`] for the `Analyze Particles` function.

After the `3D Objects Counter`, we will also obtain a "tagged map" of objects in the image (Fig. 3.8). This "tagged map" is an image with different objects labeled with different intensity values (1 for the first object, 2 for the second, etc...). When thinking about it, you will find that this nice image contains a lot of information! How could we take advantage of it so as to isolate a particular object? How could we extract the number of voxels per object? Take time to think about it, before reading further. Isolating an object from a tagged map is quite simple as all its voxels have the same intensity: simply threshold the image, using its tag value as the lower and upper thresholds. As for each object's number of voxels, a histogram operation [`Analyze -> Histogram`] should do the trick! Let's keep this in mind for the next step.

2 Circularity measures how round, or circular-shape like, the object is. In Fiji, the range of this parameter is between 0 and 1. The more roundish the object, the closer to 1 the circularity.

3 Compactness is a property that measures how bounded all points in the object are, e.g. within some fixed distance of each other, surface-area to volume ratio. In Fiji, we can find such measurements options from the downloadable plugin in [`Plugins >3D >3D Manager Options`].

Now that you have all the elements needed, here is your working plan to implement the function:

Working Plan:
1. Select the proper image
2. Display the threshold box
3. Pause the execution of the macro by displaying a dialog box asking the user to tune the threshold
4. Retrieve the threshold values
5. Make sure the proper options for 3D measurements are set
6. Run 3D-OC, using the input threshold and some size limits
7. Pack everything into a function, thinking about the proper arguments, if any, that should be entered for the process to be run

The correct code of the function is:

```
//Isolate the objects and get the characteristics on each image
function isolateObjects(minSize, image){
selectWindow (image);
run ("Threshold...");
waitForUser ("Adjust the threshold then press Ok");
getThreshold (lower, upper);
run ("Set 3D Measurements", "dots_size=5 font_size=10 redirect_to=none");
run ("3D object counter...", "threshold="+lower+ "slice=5 min.="+minSize+ " max.=100000 objects");
}
```

Again, in order to run the macro properly with this function, we need to call it in the main code body of the macro. For this step, since we need to tag objects in both channels, thus the function will be called twice. The advantage of creating user-defined function is nicely visible here: we won't have to re-type all the code from channel 1 to use it on channel 2. Functions should be called after Step 1 (3.4.2) is done, as shown in Code 3.2.

■■ The main macro till Step 2

```
//---------------------------------------
// Main macro part

//Step 1
normaliseNames();

//Step 2
isolateObjects(10, "Channel1");
isolateObjects(10, "Channel2");

//---------------------------------------
}
```

3.4.4 Step 3: Isolating the Overlapping Parts

Aim:
- Isolate object parts that are overlapping.
- Generate user-defined functions to perform the task: `getCommonParts` and `maskFromObjects` with proper arguments, if needed.

Hints:
- We have already defined objects in the previous step.
- On the tagged maps, background is tagged as 0.
- On the tagged maps, the first object is tagged with an intensity of 1, the second of 2, and so on.
- Logical operations could be applied to binary masks.

Since the `3D Object Counter` gives a tagged map, a simple thresholding could convert it to a binary image with background being 0, since object tags start at 1. As we will still need the tagged maps later on in the process, we will first duplicate them, and work on copies ([`Image > Duplicate`]).In ImageJ, the binary image after thresholding has value 0 (non-object pixels) and 255 (object pixels). Sometimes, a binary image of value 0 and 1 makes further analysis easier. Let's see an example: you want to measure a "positive" volume within a certain region of interest. Instead of thresholding and measuring the area within this ROI, in case the image has intensities being either 0 or 1, you can simply measure the sum of intensities within the ROI to obtain its area! To perform this, we could divide every pixel by this image's non-zero value, i.e. 255, using [`Process > Math > Divide`]. These would be the steps needed in the function `maskFromObjects`.

Once these steps have been applied to both tagged map, we end up with two new masks displaying pixel objects from channel1 and channel2. How to isolate the overlapping parts of objects between both channels ? One should generate a new image, where pixels are turned "ON" when both the corresponding pixels on the two channels are also "ON": this is a job for logical operators that we will now review.

In Fig. 3.9, there are two binary images A and B. We could see that in both channels, the overlapping regions have values higher than zero in both channels; while the rest of the two images have either one or both channels with zero background. Therefore if we multiply the two images, only the overlapping regions will show values higher than zero. Alternatively, we could also apply logic operations, which can be calculated faster than multiplication for computers. From the three logic operations: `AND`, `OR` and `XOR`, which is the one that we need? It is `AND`. So we can run [`Process > Image Calculator`], set the object maps of from the two channels as `Mask_Channel1` and `Mask_Channel2`, and set `AND` as `Operation`. We then rename the image with overlapping regions as "`Common_volumes`". These steps would go to the function `getCommonParts`. Think about where we should call `maskFromObjects`?

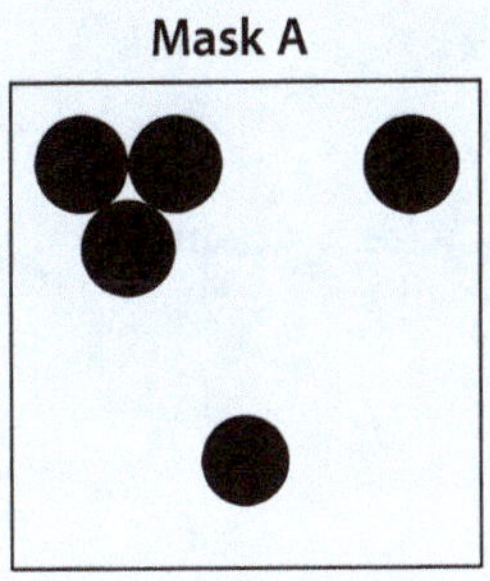

A	B	AND	OR	XOR
1	1	1	1	0
1	0	0	1	1
0	1	0	1	1
0	0	0	0	0

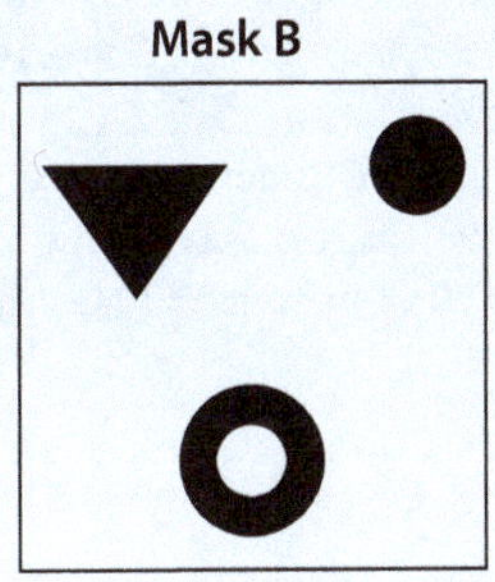

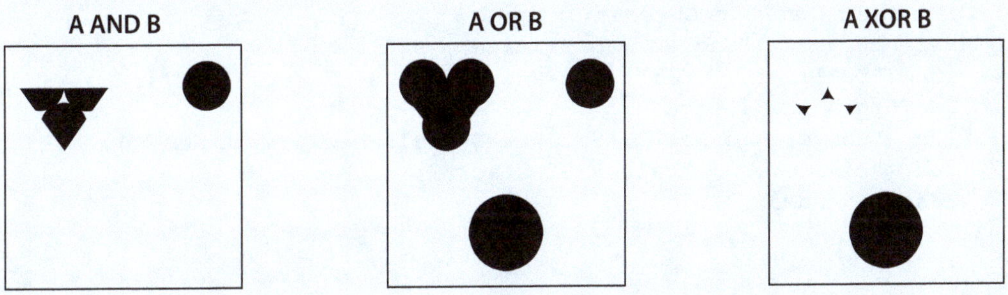

Fig. 3.9 Illustrations of applying logic operations on two binary images A and B

Working Plan:

- *Part 1: Convert an object map into a mask scaled between 0 and 1*
  1. Select the tagged map
  2. Duplicate it, giving it a name in the form "Mask_"+original name (be careful: we want the full stack, not just the current slice)
  3. Set the threshold between 1 and the maximum (65535)
  4. Convert the thresholded image to a mask
  5. Normalize the intensities between 0 and 1 (divide by 255)
  6. Pack everything into a function, thinking about the proper arguments, if any, that should be entered for the process to be run
- *Part 2: Isolate common parts from both images*
  1. Generate the normalized mask for channel 1 (i.e. 0–1 scaled)
  2. Generate the normalized mask for channel 2 (i.e. 0–1 scaled)
  3. Use logical operators between both mask to retrieve the overlapping voxels
  4. Pack everything into a function, thinking about the proper arguments, if any, that should be entered for the process to be run

The correct code of the two functions are:

```
//Generate an image of the overlapped parts from channel 1 and 2
function getCommonParts(){
//Generate the mask for channel 1
maskFromObjects("Channel1");
```

```
//Generate the mask for channel 2
maskFromObjects("Channel2");

//Combine the two masks
imageCalculator ("AND create stack", "Mask_Channel1", "Mask_Channel2");
 rename ("Common_volumes");
 }
```

```
 //Generate a mask from objects map
 function maskFromObjects(image){
selectWindow ("Tagged_map_"+image);
run ("Duplicate...", "title=Mask_"+image+ " duplicate");
setThreshold (1, 65535);
run ("Convert to Mask", "method=Default background=Dark");
run ("Divide...", "value=255 stack");
resetMinAndMax ();
}
```

And, the main code body of the macro is:

The main macro till Step 3

```
 //-----------------------------------------
 // Main macro part

 //Step 1
normaliseNames();

 //Step 2
isolateObjects(10, "Channel1");
isolateObjects(10, "Channel2");

 //Step 3
getCommonParts();
 //-----------------------------------------
}
```

3.4.5 Step 4: Retrieve Volumes

Aim:

- Measure the volumes, object by object on: `Mask_Channel1`, `Mask_Channel2` and `Common_volumes`.
- Store the volumes into arrays.
- Generate a user-defined function to perform the task: `getValues` with proper arguments, if needed.

3

Hints:

- On the tagged map, the first object is tagged with an intensity of 1, the second of 2...: the maximum intensity therefore corresponds to the number of objects in the tagged map.
- `Thresholding` with 1-1 and then `Analyze Particles` allow sending the outlines of object 1 in the `ROI Manager`.
- A non-recordable macro function exists to retrieve basic image statistics: maximum intensity, the number of pixels or the area of a ROI.

In order to find and quantify colocalised objects, we have found the overlapping (or shared) parts of the two filtered channels. We now need to identify the corresponding objects in each channel that contain these overlapping regions. To achieve this, what do we need to do? There are multiple ways, all of which involve:

1. In each channel, calculate the volume (in voxels) of each object;
2. Retrieve the volume of each overlapping region;
3. Find the labels of objects in each channel that overlap with some object in the other channel.

We will have to store multiple values related to each individual object for each channel. That won't be feasible using regular variables. Therefore we have to switch to a different structure that allows to store several values—an array. Here are some tips about how they work and how they should be used.

Technical Points: Using arrays to store multiple values

- An array is like a box, with a tag and multiple compartments
- An array should be initialized using the `newArray` keyword: either with no content but a `size`, e.g. `myArray=newArray(3);` or with a `content`, e.g. `myArray=newArray(1,50,3);`
- To attribute a content to an array, the "=" sign should be used between the compartment address between "[]" and the content, e.g. `myArray[1]=25;`
- The size of an array can be retrieved by concatenating `.length` to the name of the array, e.g. `myVariable=myArray.length;`

Now we know what is it to store data, we should think about how to retrieve them. What we want to do is to define for each object and for each channel the total volume of the object, and the volume of the object that is involved in the colocalisation process. How should we proceed? Let's think about a strategy, keeping in mind the following four important points:

1. On the tagged map image, each object is identified by its intensity (i.e. the tag): a simple thresholding from tag to tag as upper and lower threshold values allows isolating it.
2. `Analyze particles` allows exporting all object's outline to the `ROI Manager`. NB: as this function processes stacks slice-per-slice, ROIs are generated per z-plane? Depending on its spread along the z-axis, a 3D object might therefore be described by multiple 2D ROIs.

3. On the mask, each object pixel has an intensity of 1.
4. Measuring the integrated intensity within a ROI on the mask is the same as measuring the "positive" area.

Why do we decide to design such a procedure ? Let's think about what we need to retrieve. We will have to get and store the volume of all objects for channel 1 and channel 2, and the volumes involved in colocalisation for objects of both channels. All four relies on a common principle: defining the objects on a tagged map, and estimating the volume from a mask where positive pixels are labeled with a value of 1. Therefore, all four measurements can be retrieved by applying the same procedure to four different combinations of images: as the procedure is generic, building a single function is appropriate. We will call it four times, with different arguments.

Here are some technical points which should be useful for the implementation of the workflow.

Technical Points: ROI Manager-related functions

Common structure: `roiManager("function", "argument1", "argument2");`

- Some functions are recordable:
  - `Add`: push the active ROI to the ROI Manager
  - `Select`: select the i-th ROI (numbered from 0)
  - `Rename`: select the i-th ROI (numbered from 0)
- Some are not:
  - `Reset`: empty the ROI Manager
  - `Count`: returns the number of ROI within the ROI Manager
  - `Set color`: self-explanatory

To help you in this task, here is the working plan you should have come up with. Afterwards, it is always good to start writing the planned steps as comments in the macro file, and then fill with corresponding real code that you recorded and modified.

Working Plan:

1. Select the tagged map
2. Retrieve the total number of objects. Think: what is the tag of the last detected object? How to retrieve statistics from a stack ?
3. Create an array to store the volume of objects. Think: what should be the size of this array ?
4. Loop the following for each object:
   (a) Select the tagged map
   (b) Set the threshold to highlight only one object
   (c) Empty the `ROI Manager`
   (d) Run the `Analyze Particles` function
   (e) Initialize a temporary variable to store current object's volume
   (f) Loop for every found ROI:

   i. Activate the image where the quantification will be done
   ii. Activate the proper ROI
   iii. Retrieve the region's statistics
   iv. Modify the temporary variable accordingly
  (g) Push the temporary variable's content to the corresponding array compartment
5. Pack everything into a function, thinking about the proper arguments, if any, that should be entered for the process to be run, and the output that should be made by the "return" statement

Here is the correct code for the function `getValues`:

```
 //Retrieve volumes object per object
 function getValues(objectsMap, imageToQuantify){
//Activate objects' map
selectWindow (objectsMap);

//Get and store the number of objects
getStatistics (area, mean, min, nObjects, std, histogram);

//Create an output array, properly dimensioned
 measures= newArray (nObjects);

 //For each object
 for (i=1; i<=nObjects; i++){
   //Activate the objects' map
   selectWindow (objectsMap);

   //Set the threshold to select the current object
   setThreshold (i, i);

   //Empty the ROI Manager
   roiManager ("Reset");

   //Run analyze particles, add outlines to ROI Manager
   run ("Analyze Particles...", "add stack");

   //Create a variable to store the volume and initialise it to zero
   singleMeasure=0;

   //For each outline
   for (j=0; j<roiManager("Count"); j++){
     //Activate the image on which to measure
     selectWindow (imageToQuantify);

     //Select the ROI
     roiManager ("Select", j);

     //Measure the volume
     getStatistics (area, mean, min, max, std, histogram);
```

```

        //Add the volume to the variable
        singleMeasure+=area*mean;
      } //End for each outline

    //Push the measure to the output array
    measures[i-1]=singleMeasure;

  } //End for each object

  //Return the output array
  return measures;
}
```

And it will be called four times in the main macro code to calculate values for both channels:

▪▪ The main macro till Step 4

```
1    //--------------------------------------
2   // Main macro part
3
4   //...
5
6   //Step 4
7  objectsVolume1=getValues("Tagged_map_Channel1", "Mask_Channel1");
8  commonVolume1=getValues("Tagged_map_Channel1", "Common_volumes");
9
10  objectsVolume2=getValues("Tagged_map_Channel2", "Mask_Channel2");
11  commonVolume2=getValues("Tagged_map_Channel2", "Common_volumes");
12   //--------------------------------------
13  }
```

To avoid duplicates of same code, in code 3.4 line 4 represents the part in Code 3.3.

3.4.6 Step 5: Generate Outputs

Aim:

- With the measure for each channel, calculate the ratio of volume involved in colocalisation.
- Display the results in a `resultsTable`, one row per object.
- Build a map where the intensity of object corresponds to the ratio of overlap.
- Generate a user-defined function to perform the task: `generateOutputs` with proper arguments, if needed.

3

Hints:
- On the tagged map, the first object is tagged with an intensity of 1, the second of 2..: the maximum intensity therefore corresponds to the number of objects.
- Caution: the ratios are decimal values, intensities on all images we are working on so far are integers.
- In `ImageJ`, `[Process/Macro]`, there is a way to replace an intensity by another value.

Numbers have been extracted: well done! But still, the user can't see that: we need to create some outputs. Two ideas come up: should I display a table with all numbers? Or is the user rather of a visual kind and would require to have the values mapped to an image? Ok, let's not make a decision, but do both!

We will first generate a table per channel, with one row per object and three columns containing the volume of the object, the volume of this object involved in colocalisation (i.e. overlapping) and the ratio. Before starting coding this output, we should have a look at the possibilities of table output offered by ImageJ. This is the topic we will cover in the next technical point.

Technical Points: Using a `ResultsTable` to output data
- `[Analyze/Clear Results]`: empties any existing `ResultsTable`. This function is macro-recordable.
- `nResults`: predefined variable to retrieve the number of rows.
- `setResult("Column", row, value)`: Adds an entry to the ImageJ results table or modifies an existing entry. The first argument specifies a column in the table. If the specified column does not exist, it is then added. The second argument specifies the row, where `0<=row<=nResults`.
- To add a row in the table, simply address the last row+1: as the rows are numbered 0 to nResults-1, address row nResults.
- To add a column, one simply needs to give it a name.
- The rows might be labeled using "Label" as column name.

When reading the technical points, it seems that the output as a table is not that complicated. Do not forget to tag each line with a note about which object it is about! One additional trick: as two channels are analyzed, each of which should reside in a table, but only one table would be manipulated at a time. To solve this, once more, ImageJ's Built-in Macro Functions page might help: try to find a function that could rename a `ResultsTable`.

Let's deal with the output as a colocalisation map. First, we need a container to push the ratios into. The simplest way is to duplicate an already existing image. Let's take the tagged map and make a duplication. The resulting image is a 16-bits type image. This implies that only integer intensities can be stored in it. As you may guess, ratios are decimal values:

if we push them directly, the values will be clipped and we are running into the risk of obtaining a black image! Image type conversion is therefore required and should be performed using the `[Image > Type > 32-bit]` menu. Finally, how and where to put the ratio? The easiest way would be to identify the object by its tag, then replace all the values of its pixels by the ratio. This is a straightforward process when knowing about the `[Process > Batch > Macro]` function. This menu allows applying some logic to images. In our case, we will have to enter something like `if(v=tag) v=ratio;` (to be adapted with the proper values).

We now have all the elements to build the function for this part of the workflow, let's have a look at the working plan.

Working Plan:

1. Make some clean-up! Empty any existing `ResultsTable`
2. Activate the tagged map
3. Remove any existing ROI
4. Duplicate this image to generate the future colocalisation map
5. Properly convert the image
6. Loop for all objects:
   (a) Calculate the volume ratio
   (b) Push a proper label to the `ResultsTable`
   (c) Push the object's volume to the `ResultsTable`
   (d) Push the object's volume involved in the colocalisation to the `ResultsTable`
   (e) Push colocalisation ratio to the `ResultsTable`
   (f) Activate the colocalisation map
   (g) Replace the the intensity of any pixel carrying the current object's tag by the ratio
7. Pack everything into a function, thinking about the proper arguments, if any, that should be entered for the process to be run

Now that you have implemented your own version of the code, you may compare it to the functions we have implemented.

```
//Generates two types of outputs: a results table and 2 co-localisation maps
function generateOutputs(objectsMeasures, commonMeasures, objectsMap){
//Empties any pre-existing results table
run("Clear Results");

//Duplicate the objects map
selectWindow(objectsMap);
run("Select None"); //Needed to remove any ROI from the image
run("Duplicate...", "title=Coloc_Map duplicate");
 run("32-bit"); //Needed to accomodate decimal intensities

```

```
for (i=0; i<objectsMeasures.length; i++){
  //Calculate the ratio
  ratio=commonMeasures[i]/objectsMeasures[i];

  //Fill the results table with data
  setResult ("Label", nResults, "Object_"+(i+1));
  setResult ("Full object", nResults-1, objectsMeasures[i]);
  setResult ("Common part", nResults-1, commonMeasures[i]);
  setResult ("Ratio", nResults-1, ratio);

  //Replace each object's tag by the corresponding colocalisation ratio
  selectWindow ("Coloc_Map");
 run ("Macro...", "code=[if(v=="+(i+1)+ ") v="+ratio+ "] stack");
}
resetMinAndMax();
}
```

Once more, to test this new function, some lines should be added to the main body of our macro. An example is given hereafter.

■■ My main macro till Step 5

```
//----------------------------------------
// Main macro part

//...

//Step 4
objectsVolume1=getValues("Tagged_map_Channel1", "Mask_Channel1");
commonVolume1=getValues("Tagged_map_Channel1", "Common_volumes");

objectsVolume2=getValues("Tagged_map_Channel2", "Mask_Channel2");
commonVolume2=getValues("Tagged_map_Channel2", "Common_volumes");

//Step 5
generateOutputs(objectsVolume1, commonVolume1, "Tagged_map_Channel1");
IJ.renameResults("Volume_colocalisation_Channel1");
 selectWindow ("Coloc_Map");
rename ("Volume_colocalisation_Channel1");

generateOutputs(objectsVolume2, commonVolume2, "Tagged_map_Channel2");
IJ.renameResults("Volume_colocalisation_Channel2");
selectWindow ("Coloc_Map");
rename ("Volume_colocalisation_Channel2");
//----------------------------------------
```

3.4.7 Step 6: Make the Macro User Friendly

Aim:
- A graphical user interface should be displayed first, to ask the user for the parameters to use for analysis.
- This step should be implemented as a function: GUI with the proper set of argument(s), if needed.
- The function should return the entered values as an array.

Hints:
- Identify which parameters are there
- Use previous technical points.

During step 1 (▶ Sect. 3.4.2), we have had a glimpse at user interactions: we used the `waitForUser` statement to pause the execution of the macro, and to ask for a user's intervention. There is another possible interaction: Graphical User Interface (GUI). GUI are dialog boxes, which can be fed with parameters. Our first task is to review all parameters that are user-defined, then to build a proper GUI. When looking at the workflow, we can identify two such parameters: the minimum and maximum expected sizes of objects to be isolated from both channels' images. We will build a basic GUI, asking for those two parameters. We will need to learn about the instructions to be used, detailed in the next technical points.

Technical Points: Generating Graphical User Interface
- Initialise a new GUI: use `Dialog.create(''Title'')`
- Add content to the GUI, where content could be number, string, checkboxes… e.g.: `Dialog.addNumber(label, default)` adds a numerical field to the dialog, using the specified label and default value.
- Display the GUI: `Dialog.show()`
- Retrieve the values in order: one instruction that retrieves the first number, then with a new call, the second…e.g.: `Dialog.getNumber()` returns the contents of the next numeric field.

Based on the above information, the building steps are quite simple:

Working Plan:
1. Create a new Dialog Box
2. Add a numeric field to retrieve the minimum expected size of objects
3. Add a numeric field to retrieve the maximum expected size of objects

4. Display the Dialog Box
5. Create a properly sized array to store the input data
6. Fill the array with retrieved data
7. Pack everything into a function, thinking about the proper arguments, if any, that should be entered for the process to be run, and to output that should be made by the "return" statement

Now it is your turn to do some coding! Try to go ahead yourself first, before looking at our version of the implementation below!

```
//Display the graphical user interface
function GUI(){
 Dialog.create ("colocalisation");
 Dialog.addNumber ("Minimum size of objects on channel1 (in voxels)", 10);
 Dialog.addNumber ("Minimum size of objects on channel2 (in voxels)", 10);
 Dialog.show ();

 out= newArray (2);
 out[0]= Dialog.getNumber ();
 out[1]= Dialog.getNumber ();

 return out;
}
```

We will revisit Code 3.1: as parameters are now stored in an array after applying our own defined function GUI, we need to use them in the function calls.

```
 //----------------------------------------
 // Main macro part

 //Step 6
parameters=GUI();

 //Step 1
normaliseNames();

 isolateObjects(parameters[0], "Channel1");
 isolateObjects(parameters[1], "Channel2");
```

3.4.8 What Then?

What then? First, lay done, have a nice cup of tea, and get ready for a review of what you've achieved so far! The macro works nicely now and you've achieved all aims we've fixed on ◘ Table 3.1. Here is an update (◘ Table 3.2):

Table 3.2 What have we learned so far?

Step	What is it	What have learned
☑ 1	Normalize the data name (▶ Section 3.5.2)	Re-use & adapt existing code easily: thanks to user-defined functions
☑ 2	Tag the objects (▶ Section 3.5.3)	Waiting for an action from the user Using a plugin from a macro
☑ 3	Isolate the overlapping parts (▶ Section 3.5.4)	Manipulating binary masks
☑ 4	Retrieve volumes (▶ Section 3.5.5)	Exploiting the plugin's outputs, Using the ROI Manager (repeat)
☑ 5	Generate outputs (▶ Section 3.5.6)	Using the Results Table to output data (repeat), Generating output images
☑ 6	Make the macro user friendly (▶ Section 3.5.7)	Generating Graphical User Interfaces

So, are we done? Since users' requests always evolve, we won't have to wait long till the user comes back, with his mind changed or asking for more...Get ready for next challenges in ▶ Sect. 3.5!

3.5 Workflow 2: Objects Overlap Intensity Quantification

Scenario

» It works well...but...

I have now the impression that the overlap might not be the main parameter. I think what matters is the amount of protein engaged in the colocalisation process.

Therefore, I would like to quantify object per object, channel per channel the percentage of protein involved in the process and get a map of quantification.

The user comes back to the Facility...but he has changed his mind

3.5.1 What Should We Do?

The answer is quite simple. First, we will go for a loud primal scream: we spent 3 h coding a macro that most probably won't be used!? Second, we will be tempted to trash all what we've done, and heavily swear at the user who has no clue about what he really needs. Finally, we will think again at the beauty of our code, and try to find a way to re-use it.

Lucky enough, we have been structuring our code into functional blocks. Several of them can be re-used, as we only need to adapt the analysis part and the GUI. In study case 1, ▶ Sect. 3.4.5, we used a normalized mask and determined the intensities on it, using the 0–1 scaling

Table 3.3 What more do we need to do?

Step	What is it	What to implement
☑ 1	Normalize the data name (Same as in ▶ Section 3.5.2)	~~Re-use & adapt existing code~~ easily: ~~thanks to user-defined functi~~ons DONE!
☑ 2	Tag the objects (Same as in ▶ Section 3.5.3)	~~Waiting for an action from t~~he user ~~Using a plugin from a mac~~ro DONE!
☑ 3	Isolate overlapping parts (Same as in ▶ Section 3.5.4)	~~Manipulating binary masks~~ DONE!
4	**Retrieve intensities (Todo in ▶ Section 3.6.2)**	**Exploiting the plugin's outputs, Using the ROI Manager (repeat)**
☑ 5	Generate outputs (Same as in ▶ Section 3.5.6)	~~Using the ResultsTable to outp~~ut data ~~(repeat), Generating output~~ images DONE!
6	**Make it user friendly (To adapt in ▶ Section 3.6.3)**	**Generating Graphical User Interfaces**

to report for volumes. Instead of using a scaled map directly, we could try to retrieve the original intensities. This strategy needs a bit of thinking, which we'll do in ▶ Sect. 3.5.2.

Next table gives an overview of our workflow and the steps to be adapted (◘ Table 3.3).

3.5.2 New Step 4: Retrieve Intensities

Aim:
- We need to generate new inputs for `getValues` function.
- This step should be implemented as a function: `getMaskedIntensities` with the proper set of argument(s), if needed.

Hints:
- We already have generated functions to get masks within intensity of 0 for background and 1 for objects.
- Arithmetics is possible between images via `[Process >Image Calculator]`.

Now we have settled down on user's request, we realize that the workflow is actually not too much work to implement and adapt. We have already been able for each channel to determine for each object its volume, and the part of its volume involved in the co-localization process, allowing to generate a physical co-localization percentage map.

What we now need to do is exactly the same process replacing the volume per object by the total fluorescence intensity associated to it. In the previous case study, we have fed the `getValues` function with the 0–1 scaled mask and the tagged map. We now should feed it with the original image intensities, together with the tagged map. We already have the normalized mask. It can be used together with the original image to get a filtered-intensities map. The operation to be performed is a simple mathematical multiplication between both images. This is the purpose of the `[Process >Image Calculator]` function from ImageJ. This operation should be performed using both the objects' maps for a channel and for the overlaps' mask. With all these in mind, the plan is easy to form:

Working Plan:

1. Use `[Process > Image Calculator]` function between the objects' mask and the original image
2. Rename the resultant image as "Masked_Intensities"
3. Use `[Process > Image Calculator]` function between the overlaps' mask and the original image
4. Rename the resultant image as "Masked_Intensities_Common"
5. Pack everything into a function, thinking about the proper arguments, if any, that should be entered for the process to be run

And its implementation takes the following form:

```
//Generate masked intensities images for the input image
function getMaskedIntensities(mask, intensities){
imageCalculator ("Multiply create stack", mask,intensities);
rename ("Masked_intensities_"+intensities);
imageCalculator ("Multiply create stack", "Common_volumes",
intensities);
rename ("Masked_intensities_Common_"+intensities);
}
```

To be able to launch and test the new version of the macro, we will replace Code 3.5 by:

```
//---------------------------------------
// Main macro part

//...

//Step 4
getMaskedIntensities("Mask_Channel1", "Channel1");
getMaskedIntensities("Mask_Channel2", "Channel2");

objectsIntensity1=getValues("Tagged_map_Channel1", "Masked_inten-
sities_Channel1");
commonIntensity1=getValues("Tagged_map_Channel1", "Masked_inten-
sities_Common_Channel1");

```

```
objectsIntensity2=getValues("Tagged_map_Channel2", "Masked_intensities_Channel2");
commonIntensity2=getValues("Tagged_map_Channel2", "Masked_intensities_Common_Channel2");

 //Step 5
generateOutputs(objectsIntensity1, commonIntensity1, "Tagged_map_Channel1");
IJ.renameResults("Intensity_colocalisation_Channel1");
 selectWindow ("Coloc_Map");
rename ("Intensity_colocalisation_Channel1");

generateOutputs(objectsIntensity2, commonIntensity2, "Tagged_map_Channel2");
IJ.renameResults("Intensity_colocalisation_Channel2");
selectWindow ("Coloc_Map");
rename ("Intensity_colocalisation_Channel2");
//----------------------------------------
```

3.5.3 Adapted Step 6: Make the Macro User Friendly

Aim:

- Modify the GUI function to add choice between the two types of analysis.
- The GUI function should return an array containing the 2 former parameters, plus the choice made by the user.

Hints:

- We already have all the required code.
- We previously have seen some function exist to create GUIs: have a look at IJ build-in macro functions webpage.
- We previousy have seen that structures exist to branch actions on users' input.

During the previous step (▶ Sect. 3.5.2, we have introduced a new way to process the images. This is a second option to interpret the colocalisation: rather than being based on the physical overlap, it deals with the distribution of molecules, as reported by the associated signal. Both options could be used in colocalisation studies, and it would be ideal to be able to switch from one option to another, e.g. by choosing it from a drop-down menu.

We already have seen how to create a GUI. The remaining step consists of customizing the existing one so that it accommodates the list and behaves according to the user choices. Therefore, surely the required instructions would start with something like `Dialog.addSomething` and `Dialog.getSomething`. Refer to ImageJ's Built-in Macro Functions page to find the proper instructions.

Once the parameters are set, the behavior of the macro should adapt. We should use a conditional operation of the different parts of the macro, with a proper structure that is described in the next technical point.

Technical Points: Generating Graphical User Interface

- Several types of instructions exist to branch the execution of certain part of the code according to a parameter' value. For example, a parameter can be tested against one reference value, against two or more.
- `if(condition) {}` statement: a boolean argument is provided to the `if` statement. If `true`, the block of instructions reside between the curly brackets are executed. If not, the macro will resume its execution, starting with the first line after the closing bracket. The argument generally takes the form of two members, separated by a comparison operator. Amongst operators, greater than sign, >, or lower than sign, <, could be used. Greater/lower than or equal to statement could be generated by addition the equal sign to the formers: =. As for equality, the comparison statement is formed by doubling the equal sign ==, to be not confused with the attribution statement (simple sign). Finally, the non equality might be checked either by <> or by constructing a "not" in front of the "equal" statement: ! =.
- `if(condition) {} else{}` statement: same as above, except this structure also specifies a way to handle the alternative behavior, in case the argument is false.
- `if(condition1) {} elseif(condition2) {} ...elseif (conditionN) {}` statement: this structure allows testing several conditions and adapt accordingly. This is an alternative way to achieve same result as the `switch/case` statement, which is not handled by the ImageJ macro language.

And here is the working plan of the final building block:

Working Plan:

1. Create a new array containing the possible options for colocalisation analysis methods
2. Modify the Dialog Box to include a drop-down list allowing selection of colocalisation methods
3. Retrieve the colocalisation method chosen by the user
4. Modify the return statement to take into account the new parameter
5. Adapt the behavior of the main part of the macro, depending on the user's choice

The adapted GUI function now looks like this:

```
 //Display the graphical user interface
 function GUI(){
items= newArray ("Intensity", "Volume");

Dialog.create("colocalisation");
Dialog.addNumber ("Minimum size of objects on channel1 (in vox-
els)", 10);
```

```
Dialog.addNumber ("Minimum size of objects on channel2 (in voxels)", 10);
Dialog.addChoice ("Analysis based on", items);
Dialog.show ();

 out= newArray (3);
 out[0]= Dialog.getNumber ();
 out[1]= Dialog.getNumber ();
   //Same kind of elements should be stored in an array
 out[2]=0;
   //The returned string is encoded as a number
 if (Dialog.getChoice ()== "Volume") out[2]=1;
 return out;
 }
```

A possible implementation of our macro's main body could be implemented as follows:

```
 if (parameters[2]==0){
//Performs intensity-based analysis

getMaskedIntensities("Mask_Channel1", "Channel1");
getMaskedIntensities("Mask_Channel2", "Channel2");

objectsIntensity1=getValues("Tagged_map_Channel1", "Masked_intensities_Channel1");
commonIntensity1=getValues("Tagged_map_Channel1", "Masked_intensities_Common_Channel1");

 objectsIntensity2=getValues("Tagged_map_Channel2", "Masked_intensities_Channel2");
 commonIntensity2=getValues("Tagged_map_Channel2", "Masked_intensities_Common_Channel2");

 generateOutputs(objectsIntensity1, commonIntensity1, "Tagged_map_Channel1");
 IJ.renameResults("Intensity_Colocalisation_Channel1");
 selectWindow ("Coloc_Map");
 rename ("Intensity_Colocalisation_Channel1");

 generateOutputs(objectsIntensity2, commonIntensity2, "Tagged_map_Channel2");
 IJ.renameResults("Intensity_Colocalisation_Channel2");
 selectWindow ("Coloc_Map");
 rename ("Intensity_Colocalisation_Channel2");
} else {
 //Performs volume-based analysis
 objectsVolume1=getValues("Tagged_map_Channel1","Mask_Channel1");
 commonVolume1=getValues("Tagged_map_Channel1", "Common_volumes");

 objectsVolume2=getValues("Tagged_map_Channel2", "Mask_Channel2");
 commonVolume2=getValues("Tagged_map_Channel2", "Common_volumes");

```

```
generateOutputs(objectsVolume1, commonVolume1, "Tagged_map_Channel1");
IJ.renameResults("Volume_Colocalisation_Channel1");
selectWindow ("Coloc_Map");
rename ("Volume_Colocalisation_Channel1");

generateOutputs(objectsVolume2, commonVolume2, "Tagged_map_Channel2");
IJ.renameResults("Volume_Colocalisation_Channel2");
selectWindow ("Coloc_Map");
rename ("Volume_Colocalisation_Channel2");
}
```

Take Home Message

Thanks to our user's uncertainty (and to our patience), we have come up with a flexible macro that performs both object-based and intensity-based colocalisation analysis. During the process of implementation, we have applied strategies that allow us to format the code in such a way that is easy to adapt and modify the overall workflow. The use of functional blocks is the key element in adapting the behavior of our macro.

For both methods, we have generated quality control images in the forms of colocalisation maps. A simple table as an output might be difficult to visually interpret or to link to positional clues on the image. With this type of map, the user can visually inspect the output of our macro, and adapt its parameters to make the analysis even more accurate (Table 3.4).

Table 3.4 Overall, what have we learned?

Step	What is it	What have learned
1	Normalize the data name (▶ Section 3.5.2)	Re-use & adapt existing code easily: thanks to user-defined functions
2	Tag the objects (▶ Section 3.5.3)	Waiting for an action from the user Usinga plugin from a macro
3	Isolate the overlapping parts (▶ Section 3.5.4)	Manipulating binary masks
4	Retrieve volumes (▶ Section 3.5.5)	Exploiting the plugin's outputs, Using the ROI Manager (repeat), Adapt existing code
5	Generate outputs (▶ Section 3.5.6)	Using the ResultsTable to output data (repeat), Generating output images
6	Make the macro user friendly (▶ Section 3.5.7)	Generating Graphical User Interfaces Branch on user's input

Acknowledgements We are very grateful to Anna Klemm (BioImage Informatics Facility, SciLifeLab, Uppsala University) for reviewing our chapter and suggesting extremely relevant enhancements to the original manuscript. In particular, the workflow exposition and clarification made through Fig. 3.7 results from one of her suggestions.

3

Bibliography

Bolte S, Cordelières FP (2006) A guided tour into subcellular colocalization analysis in light microscopy. J Microsc 224:213–232

Cordelières FP, Bolte S (2014) Experimenters' guide to colocalization studies: finding a way through indicators and quantifiers, in practice. Methods Cell Biol 123:395–408

Dunn KW, Kamocka MM, McDonald JH (2011) A practical guide to evaluating colocalization in biological microscopy. Am J Physiol Cell Physiol 300:C723–C742

Fletcher PA, Scriven DR, Schulson MN, Moore DW (2010) Multi-image colocalization and its statistical significance. Biophys J 99:1996–2005

Lagache T, Meas-Yedid V, Olivo-Marin JC (2013) A statistical analysis of spatial colocalization using Ripley's K function. In: ISBI, pp 896–899

Obara B, Jabeen A, Fernandez N, Laissue PP (2013) A novel method for quantified, superresolved, three-dimensional colocalisation of isotropic, fluorescent particles. Histochem Cell Biol 139(3):391–402

Rizk A, Paul G, Incardona P, Bugarski M, Mansouri M, Niemann A, Ziegler U, Berger P, Sbalzarini IF (2014) Segmentation and quantification of subcellular structures in fluorescence microscopy images using Squassh. Nat Protoc 9(3):586–596

Sage D, Donati L, Soulez F, Fortun D, Schmit G, Seitz A, Guiet R, Vonesch C, Unser M (2017) Deconvolutionlab2: an open-source software for deconvolution microscopy. Methods 115:28–41

Wörz S, Sander P, Pfannmöller M, Rieker RJ, Joos S, Mechtersheimer G, Boukamp P, Lichter P, Rohr K (2010) 3D geometry-based quantification of colocalizations in multichannel 3D microscopy images of human soft tissue tumors. IEEE Trans Med Imaging 29(8):1474–1484

Zinchuk V, Zinchuk O (2008) Quantitative colocalization analysis of confocal fluorescence microscopy images. Curr Protoc Cell Biol U4:19

The NEMO Dots Assembly: Single-Particle Tracking and Analysis

Jean-Yves Tinevez and Sébastien Herbert

K. Miura, N. Sladoje (eds.), *Bioimage Data Analysis Workflows*, Learning Materials in Biosciences,
https://doi.org/10.1007/978-3-030-22386-1_4

What You Learn from This Chapter

The aim of this chapter is to learn the principles and pitfalls of single-particle tracking (SPT). Tracking in general is very important for dynamic studies, as it is about propagating object identities over time, permitting the calculation of dynamic quantities such as object velocities. Tracking is often the first step in analyzing dynamics.

The output of tracking is simply tracks, and later steps involve computing relevant quantities from these tracks. In the case of the applications we use in this module, we wanted to learn something about the particles we track, which are unknown organelles (at the time of the publication) appearing transiently in cells upon stimulation by an interleukin. Namely we want to determine whether they are bound to a structure, transported or freely diffusing. To do so, the analysis is completed by performing a Mean Squared-Displacement (MSD) analysis.

4.1 Introduction

The data and analysis we will perform in this module is taken from the following paper: *TNF and IL-1 exhibit distinct ubiquitin requirements for inducing NEMO-IKK supramolecular structures*, Tarantino et al. (2014). In particular, we will reproduce the analysis of the paragraph "NEMO-containing punctae are slow-moving anchored structures located close to the cell surface".

Nuclear factor KB (NF-KB) essential modulator (NEMO) is a regulatory component of the IKB kinase (IKK) complex and controls NF-KB activation through its interaction with ubiquitin chains. The work of Emmanuel Laplantine focuses on the mechanics of NF-KB regulation by ubiquitination. Recently, his lab showed that NEMO, a component of the IKK complex, is crucial for NF-KB activation and the linear ubiquitination by K64. Patients with a deficiency in the linear ubiquitination machinery enabled to correlate their symptoms with a defect in NF-KB activation by cytokines. This project aims at investigating the details of the NF-KB activation initiated by stimulation by IL-1 and TNF.

We engineered cells that were expressing constitutively NEMO-eGFP from a NEMO-deficient human fibroblast cell line. They allowed us to follow NEMO dynamics using high-resolution microscopy. The most marking result of the project is that stimulation with interleukin-1 (IL-1) and TNF induces a rapid and transient recruitment of NEMO into punctate structures. These structures appear briskly, probably assembling from a pool of NEMO soluble in the cytoplasm, and roughly constant in quantity in the cell over time. They disassemble in, on average, 15–45 min depending on the stimulus (TNF or IL-1).

Our part of the project revolved around the characterization of these previously unknown structures via imaging and image analysis, completing data obtained by biochemistry. These structures and their dynamics proved to be extremely sensitive to light, and their study required careful imaging with a dedicated protocol. The part we will cover in this module focuses on a single question: Are these punctate structures bound to the membrane, freely diffusing or transported?

This simple question was important to backup the biochemistry data. Since the punctate structures were not described before and their function unknown, their motion type could give us clues about their function. For instance, if they are transported, they may be internalized in some vesicles and transported from membrane to nucleus to convey the cell activation signal. If they are freely diffusing, they may be supramolecular structures polymerizing upon some signal.

Our first imaging protocol falsely led us to think that they were transported towards the nucleus after assembly. However, these movements proved to be artifacts and caused by a phototoxic effect.[1] A second imaging protocol involved the use of TIRF microscopy with a low illumination power, which diminished the phototoxic effects. We filmed the dots for long times and process the acquired structures to analyze their motility.

We tracked the dots in Fiji using TrackMate (Tinevez et al. 2017) and, because the dots are well separated, the tracking proved relatively easy. We then analyzed the tracks using MSD analysis, to conclude on their motility with certainty. The MSD analysis is also the subject of this module, and we will then go from Fiji to MATLAB to perform it.

This particular analysis proved that the NEMO dots are anchored, both when stimulated by IL-1 and TNF. We concluded further that they are not anchored to actin filaments or microtubules (MTs), as repeating the analysis with drugs that depolymerize the cytoskeleton did not show any change. Additional analysis showed that they were most likely anchored to the cell membrane, and that NEMO molecules were under a rapid turnover in these punctate structures. So they probably play the role of phosphorylation factories, assembled and anchored at the cell membrane, that would process quickly a large amount of the otherwise soluble NEMO proteins.

4.2 Datasets

The data for this module consists in a subset of the data only from the paper. It only features 5 movies over 2 conditions:

- Ctrl: A couple of dots can be seen wandering in the cells, even if there is not stimulation. They are permanent instead of transient, and probably non-specific. They give a control of how spurious particles would perturb our measurements.
- IL1: Following IL-1 stimulation. In the study, this was the "easy" case, for the dots were bright and large compared to *e.g.* the dots we see after TNF stimulation.

You can find it on Zenodo:

▶ https://doi.org/10.5281/zenodo.1341987. The dataset (download size about 800 MB) is organized as follow:

```
Tracking-NEMO-movies_subset
   NEMO-Ctrl
     Cell_01.tif                35.2 MB
      Cell_01.xml                5.9 MB
      Cell_01_Tracks.xml        16.6 kB
      Cell_02.tif              175.5 MB
      Cell_02.xml                5.6 MB
      Cell_02_Tracks.xml       170.4 kB
   NEMO-IL1
      Cell_02.tif               91.9 MB
      Cell_02.xml                6.4 MB
      Cell_03.tif              444.8 MB
      Cell_03.xml               53.0 MB
```

1 We documented this phototoxic effect later in Tinevez et al. (2017).

```
Cell_03_Tracks.xml        1.2 MB
Cell_04.tif             251.6 MB
Cell_04.xml              50.8 MB
Cell_04_Tracks.xml        4.6 MB
```

4

The movies themselves are not very pretty. Bright dots can be seen over a cell background caused by the soluble pool of NEMO. They bleach over time. The temporal resolution is not very high (0.5 s) and the SNR is not high either since we had to compromise on laser power to avoid cell death.

Files are `.tif` movies, made for ImageJ, with the right spatial and temporal calibration. They are already split cell by cell, and have a ROI that encloses the cell. There also are `.xml` files from TrackMate and `_Tracks.xml` files generated from TrackMate, ready to be imported in MATLAB. But we will do the tracking ourselves.

4.3 Tools and Prerequisites

- Fiji
  - Download URL: ▶ https://imagej.net/Fiji/Downloads
    It does not require any extra, as TrackMate is included in the core of Fiji.
- MATLAB
  - We rely on MATLAB for the MSD analysis part, with the Curve Fitting toolbox.
  - You need to know at least a little bit about MATLAB features, like logical indexing and structures. We will not be introducing them here.
  - Because we will install specialized functions and classes in MATLAB, you also need to know at least a little bit about the MATLAB path. ▶ https://mathworks.com/help/matlab/matlab_env/what-is-the-matlab-search-path.html
  - We built a special class to perform the analysis that you can download here: ▶ https://github.com/tinevez/msdanalyzer/zipball/master

`@msdanalyzer` is a MATLAB *class*, so you have to put the `@msdanalyzer` *folder* in the MATLAB path, but not its content.[2] For instance on my MATLAB installation, I have a folder called `/Users/tinevez/Development/Matlab/msdanalyzer` that is on the path and that contains the `@msdanalyzer` folder. But the `@msdanalyzer` folder is not in the path.

4.4 Workflow

We will deal separately with single-particle tracking in Fiji using TrackMate, and track motility analysis in MATLAB using `@msdanalyzer`. The two following sections are largely independent and present different concepts. To perform the MSD analysis, please use the dataset linked above that include them.

2 This is explained on The Mathworks website: ▶ https://mathworks.com/help/matlab/matlab_oop/organizing-classes-in-folders.html.

4.5 Single-Particle Tracking with TrackMate

TrackMate is a Fiji plugin dedicated to tracking. It can do cell-lineaging (and was initially developed for this very purpose, see Tinevez et al. (2012)) but also has automated analysis tools to perform single-particle tracking of sub-cellular structures. It ships a user-friendly graphical user interface that allows to inspect tracking results and curate them. The following part describes how to use TrackMate to generate the tracks over one of our movies. An extended documentation for this plugin can be found here: ▶ https://imagej.net/TrackMate, along with supplementary material of the associated publication.

4.5.1 Step 1: Loading Image Data and Launching TrackMate

For the example below, we will use the `Cell_02.tif` in the `NEMO-IL1` folder. You can display a `.tif` file by performing a simple drag and drop on the Fiji toolbar. This movie does not have many dots, which will simplify getting familiar with the workflow.

It is very important that you check the dimensionality of the image at this point of the analysis and correct it if required. To do so, check the image properties in Fiji ([Image] » [Properties...] menu item or [⇧]+[P], ◘ Fig. 4.1). In our case we have a 2D over time acquisition, so make sure the metadata reports 1 z-slice and 307 frames. Also, TrackMate reports any quantities (space and time) in physical units, so the pixel size

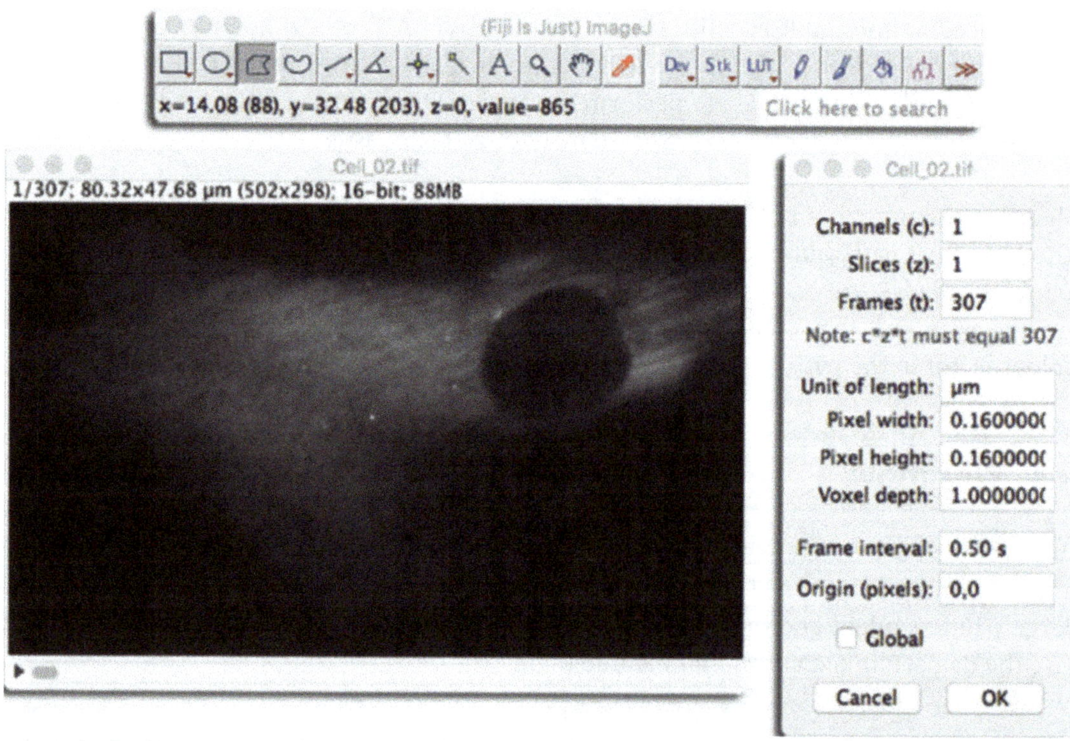

◘ **Fig. 4.1** Image properties

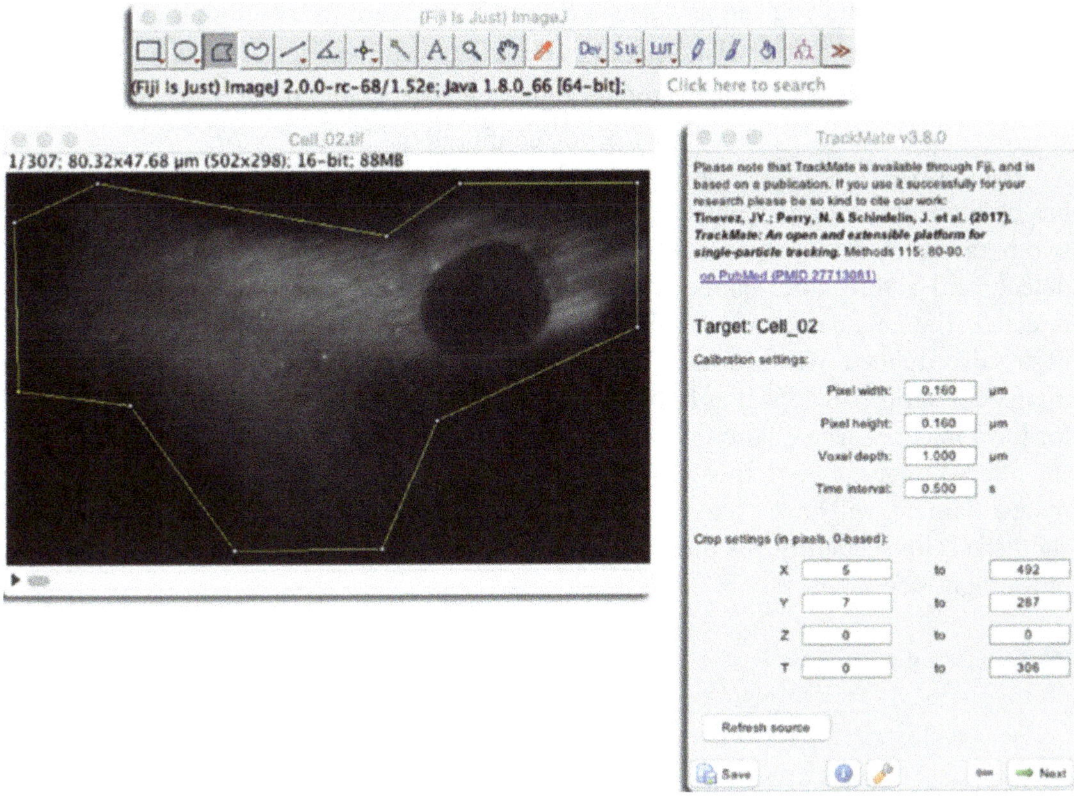

Fig. 4.2 The first panel in TrackMate UI

and frame interval must be correct since you will not be able to change them further in the analysis. For these movies, the pixel size is 0.160 µm per pixel and the frame interval is 0.5 s.

When this is done, close the properties window, make sure the active image is the one with our NEMO-labelled cell, and launch TrackMate. The plugin can be found in the Plugins › Tracking › TrackMate menu. The GUI will show up and the first panel will display a recapitulation of the image metadata. At this step, you can define a ROI that will be used for the analysis. In our case it does not matter, since the images are cropped around a cell. If you want to use a ROI in TrackMate, draw a ROI over the active image, and press the Refresh source button on the first panel. You should see the bounds changing on the panel (Fig. 4.2).

4.5.2 Step 2: Detection

In our case, the objects we want to track are NEMO dots; since they are smaller than the resolution limit of the microscope, we cannot resolve their shape, hence *segmenting* them would not bring any information that would allow us to discriminate them. They all look the same in the eye of our microscope. We need a simple *detection* algorithms that will yield their position nothing more, which is exactly what TrackMate ships.

The TrackMate user interface is inspired by the Bitplane Imaris wizard. You will find at the bottom right of the panel a Next button that will bring you to the next panel when you are done with the current one. Typically, you deal with one group of parameters or choice per panel and you can navigate backward if you want to try another one. Click on the → Next button and you will now see a panel where you can choose the detector we will use. Pick the LoG detector, which is the default, then click again on the → Next button. You are now presented with a panel that lets you configure the LoG detector.

The LoG detector is performing remarkably well for its simplicity. It excels at finding bright, blob-like, roundish objects, that we will call spots or detections and only requires two parameters: the approximate diameter (in physical units) of the objects we want to detect, and a threshold value on a quality metrics, below which detections will be rejected. The LoG detector works by filtering the image with the Laplacian of Gaussian filter (also dubbed Mexican hat filter) tuned to the specified diameter. In the filtered image, the spots will appear as bright and sharp peaks, and they are detected by looking for local maxima. The quality of a detection is the value at the local maxima location in the filtered image. Due to this image filtering step, spots smaller or larger than the estimated diameter will have a lower quality value than spots of the same intensity but of the right size. Consequently, the quality of detection is highest when the spot is bright and of the right size.

Exercise 4.1

Play with the Estimated blob diameter and Treshold parameters to find settings that would detect all NEMO dots and a limited number of spurious detections. The Preview button will show you what the chosen parameters do on the current frame only. The other parameters can be ignored.

Solution

Using a diameter of 0.5 μm and a threshold of 1000 seems OK (Fig. 4.3). We still have many spurious spots but we can filter them out later.

Click on the `Next` button to run the analysis on the whole movie. It should not take too long (a few seconds) and you should have about 9000 spots in total.

4.5.3 Step 3: Filtering

It is very important to filter out as many spurious spots as we can because detection might very well yield a large amount of them. By setting the quality threshold to a non-zero value we already filtered them a first time. When clicking → Next after the detection is completed, you are presented with the initial filtering panel (Fig. 4.4). It shows the quality histogram and allows for discarding spots with a quality lower than what we set here. In simple cases, we expect a bi-modal histogram, with two peaks well separated between spurious spots with low quality and spots with a high quality. This is not the case here, so we keep the value unchanged and move to the next panel by clicking → Next.

TrackMate lets you pick the view to display the detection and tracking results. As of today, there is only one working consistently, the HyperStack Displayer. It simply displays the results on the ImageJ hyperstack, so select this one. The spots are now displayed on the image as magenta circles (Fig. 4.5). A quick inspection reveals that we still have spurious spots. You can see them appearing and disappearing as you move in time, while true spots tend to persist over several frames.

4

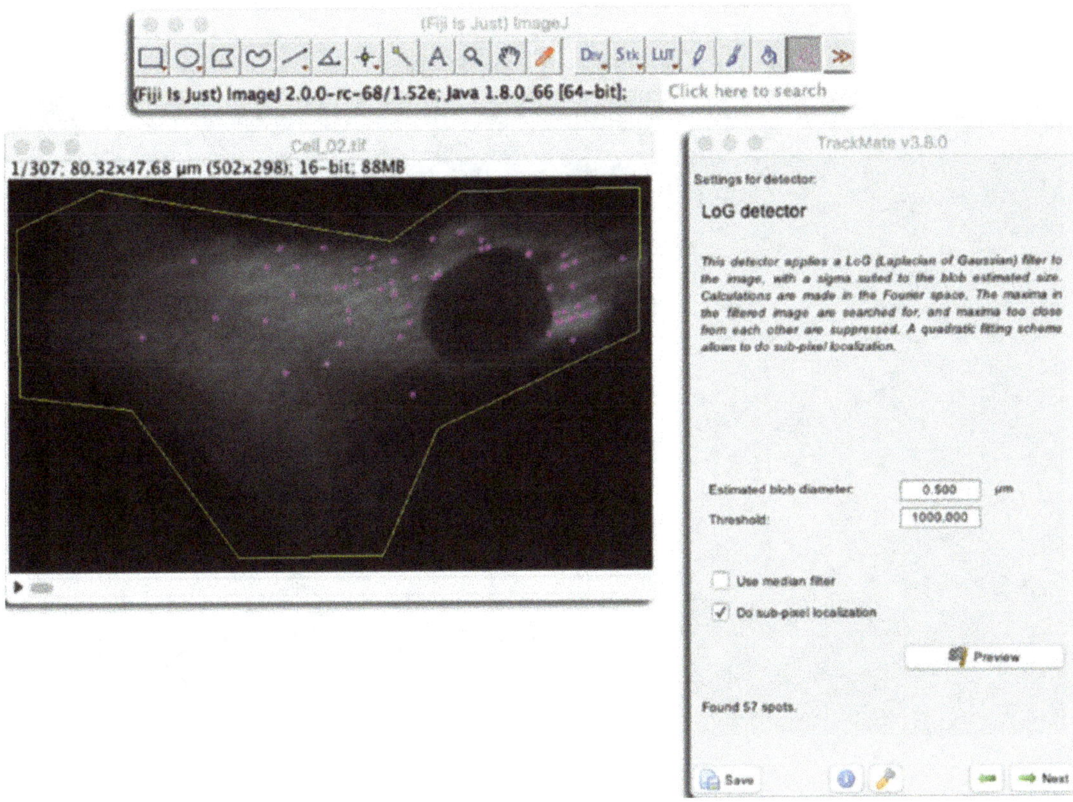

Fig. 4.3 LoG detector parameters

Fig. 4.4 Initial filtering panel

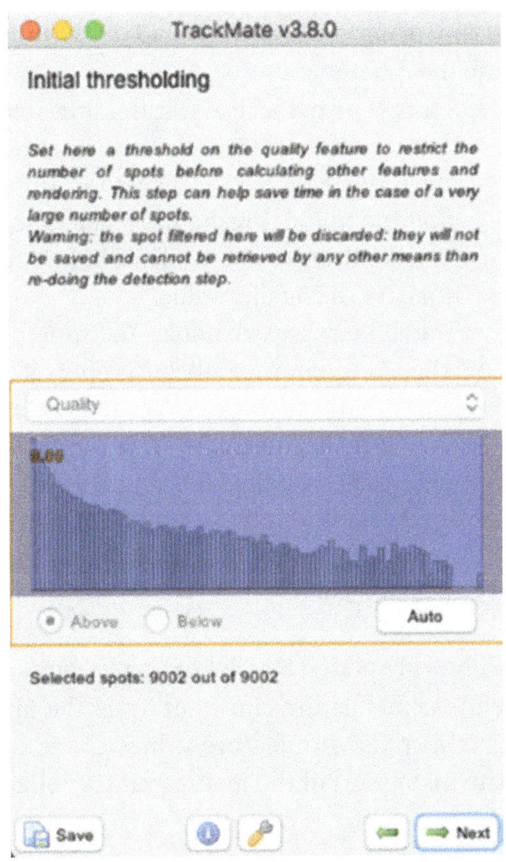

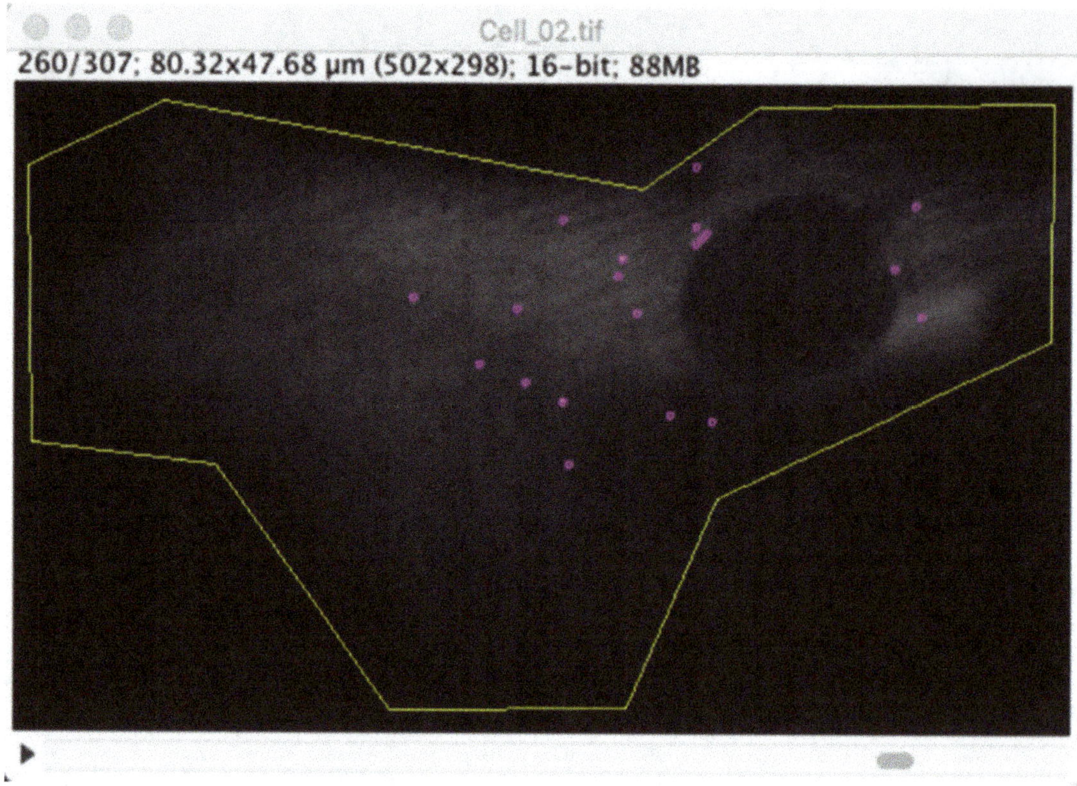

Fig. 4.5 The HyperStack Displayer with detection results

Move on to the next panel by clicking [Next]. We are now presented with another filtering panel. As we already had one before (the initial filtering panel), it is worth mentioning the differences.

1. Spots have numerical features attached to them. A feature is a numerical scalar value that reports a quantitative information on the object it represents. For instance, the mean intensity around a spot location is a numerical feature. Features are calculated *after* the initial thresholding step. The filters you set in this panel are based on features that are not available before this step and the initial filtering can only be done on the quality value.
2. These filters are *reversible*. The spots are not deleted from the data, but hidden. So when you remove a filter, the spots it discarded reappear. This is useful if you realize later that the filters were inadequate and too stringent and preventing proper linking. To adjust the filters later, you can navigate back to this panel by pressing the left green arrow on the GUI. In contrary, when using the initial thresholding, the spots are deleted, which is useful to save memory but is irreversible. If you want to retrieve the spots you discarded in the initial thresholding step, you must re-run the detection step.

Filters are added by clicking on the green button. A small panel appears that lets you choose the feature you want to set the filter on, with what value and whether we should retain spots with feature values above or below the threshold. The panel also displays the histogram of the feature values collected for all spots, and has an `Auto` button that

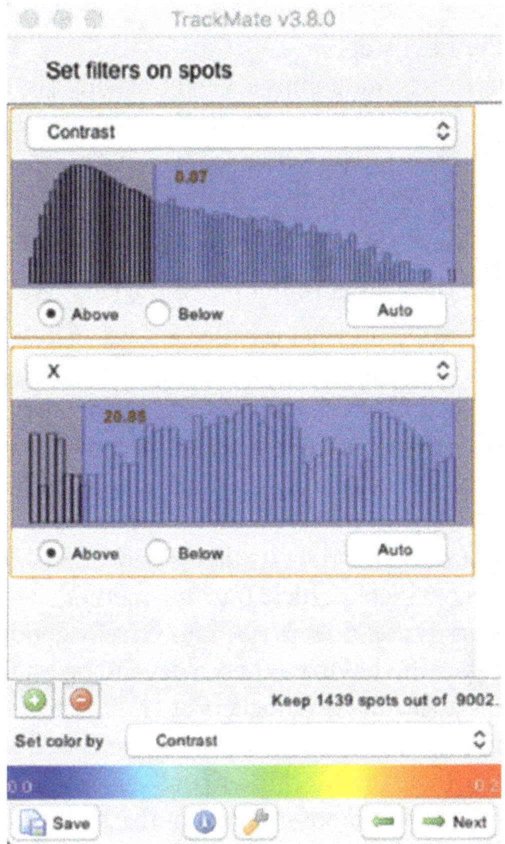

Fig. 4.6 Filtering panel with two filters on contrast and X position, additionally setting the spot color by contrast value

automatically determines a threshold value using Otsu method (Fig. 4.6). The red button removes the last filter. Also note that you can use the feature values to color the spots, using the drop-down list at the bottom of the filtering panel. Filters are combined by taking the intersection of the spots they yield. On Fig. 4.6 for instance, we retain the spots that have contrast values above 0.07 **and** whose X position are above 20.85 µm.

Exercise 4.2

Find a combination of filters that remove all spurious spots and keep the true ones.

Solution

Such a combination is hard if not impossible to find in our case. Thankfully, we do not mind a few spurious spots as we will be analyzing tracks. We will later filter out tracks made of spurious spots.

Remove all filters, set the color-by option to Uniform color and click the → Next button.

4.5.4 Step 4: Particle-Linking

Now that we have spots, we want to link them over time and build tracks. The tracks will be what we will analyze in the section dedicated to motion analysis, and we will do it in MATLAB. But for now we need to generate these tracks. In TrackMate, particle-linking happens similarly to the detection step. You are now presented with a panel that lets you select the particle-linking algorithm (or "tracker") to be used for the next step.

TrackMate ships several trackers, but the most useful ones fall in two main categories:

- The LAP-based trackers. LAP stands for Linear Assignment Problems. There are two trackers named Simple LAP tracker and LAP tracker that implement a stripped version of an algorithm published by Jaqaman et al. (2008). They are configured to deal well with objects that diffuse or move randomly.
- The Kalman-filter based trackers. We have only one, called Linear motion LAP tracker. It is based on what is called a Kalman-filter[3] introduced in the 1960s by R.E. Kalman (Kalman (1960)). Our implementation is well suited to particles that have a nearly-constant velocity vector. That is: particles that move by roughly the same amount between each frame and do not change direction too fast. Of course it can accommodate some changes of velocity provided they are modest.

Choosing the right tracker is critical. In Chenouard et al. (2014), the performance of 14 different single-particle tracking methods were assessed. One of the main conclusions of this work is that there is not a universally good tracker for all bio-imaging problems. A tracking algorithm has to be chosen depending on the motion model of the objects to be tracked. For instance, the LAP trackers of TrackMate are well suited for objects that are freely diffusing or bound. The Linear motion tracker is well suited for objects that are transported.

This causes a chicken-and-egg problem in our case, since we actually want to determine what is the motion model of the NEMO dots by analyzing tracks. Practically, we carefully looked at the movies and and assessed whether it was plausible for the dots to be transported. Their motion seemed erratic, and so we started with the LAP tracker. As we will see later, we found that the dots have a motion type for which the LAP trackers are well suited, so our choice appears valid *a posteriori*. This is close to having a circular reasoning fallacy. However, we must temper this criticism. The choice of the right tracker is important to yield accurate tracks that faithfully follow the true particles over time. The analysis of tracks is a subsequent step. So first, an inadequate choice of a tracker can be detected by checking the tracks manually, following a dozen of them and looking for jumps to another particles or early breaks. And second, we can be in a situation where the particle density and the detection quality is such that the choice of the tracker will not matter. This is the subject of an exercise below.

In TrackMate, the trackers suited for non-transported motion are the LAP trackers. They are based on minimizing the total cost to link a set of spots in one frame to the spot in the next frame, or the cost to link track segments together. We have the LAP tracker and its Simple LAP tracker version. They actually wrap the same algorithm, but the latter offers fewer configuration options. The LAP tracker can be configured to generate tracks that are splitting (as for cell division) or merging. The cost to link one spot to another one can be altered by differences in spots features values, such as intensity, radius, … The Simple LAP tracker only offers to bridge gaps in tracks caused by missed detections, and the linking costs are simply based on distance. It results in tracks being *linear*, that is without merges or splits, and at most one spot per frame. We observe that the NEMO dots do not merge or split, so this is the right tracker to pick.

3 Kalman filter on Wikipedia: ▶ https://en.wikipedia.org/wiki/Kalman_filter.

4

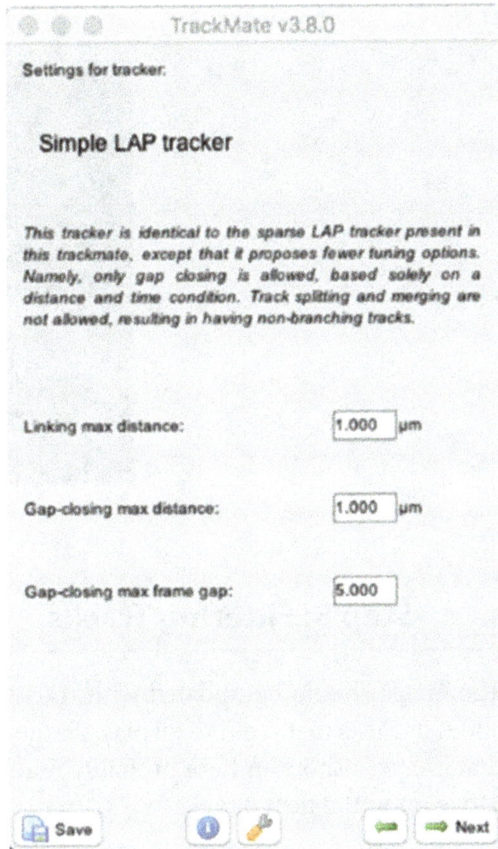

Fig. 4.7 Configuring the Simple LAP tracker

Exercise 4.3

The `Simple LAP tracker` has 3 parameters to configure it, that set

- the maximal distance to link from one frame to the next;
- the maximal frame gap to bridge over missing detections;
- the maximal distance to bridge over missing detections.

Try and find a suitable parameter set that yields acceptable results, based on you checking the tracks.

Solution

As explained in Jaqaman et al. (2008), this tracker performs spatially global optimization, and therefore is rather robust against a lot of variation in parameter values. You should find acceptable results for a wide range of parameters, provided they are not aberrant. Try some of them and click the Next button to get the tracking results displayed on the image. Click on the Previous button to change parameters and start again. Check Fig. 4.7 for values that work.

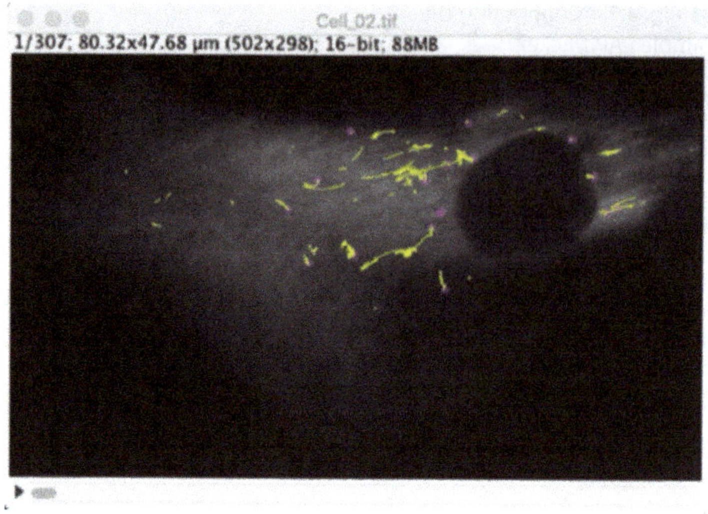

Fig. 4.8 Raw tracking results

4.5.5 Step 5: Filtering Tracks

The image should be updated with tracks, as shown in Fig. 4.8. We note that there are about 3 tracks that seem to display a large excursion in the cell. Also, there are many tracks that are very short and are probably made of spurious spots. We want to filter them out. Move on to the next panel.

Exercise 4.4

The track filtering panel works like for the spot filtering panel. Find a combination of filters that can remove tracks originating from spurious spots.

Solution

Spurious spots arise from noise in the image. As long as there are only a few of them, it is therefore very unlikely that spurious spots appear many times consecutively in the vicinity of the same location. If they are ever linked in a track, it will be short compared to tracks originating from true spots, that can be followed over many frames. A good strategy is therefore to filter tracks based on the number of spots they contain. It has a side benefit: the MSD analysis we will perform later in this chapter requires the tracks to be long for accuracy. Since the histogram for the number of spots in tracks have a large peak at low values that precede a gap at $N=20$, we can use this value as a threshold (Fig. 4.9).

So we are now left with a small number of long tracks (Fig. 4.10).

Exercise 4.5

Go back 3 steps and try to perform tracking with the Linear motion LAP tracker. As we said, it is not the optimal tracker for the motion type we suspect the NEMO dots have. Check whether the tracking results differ from those of the LAP tracker.

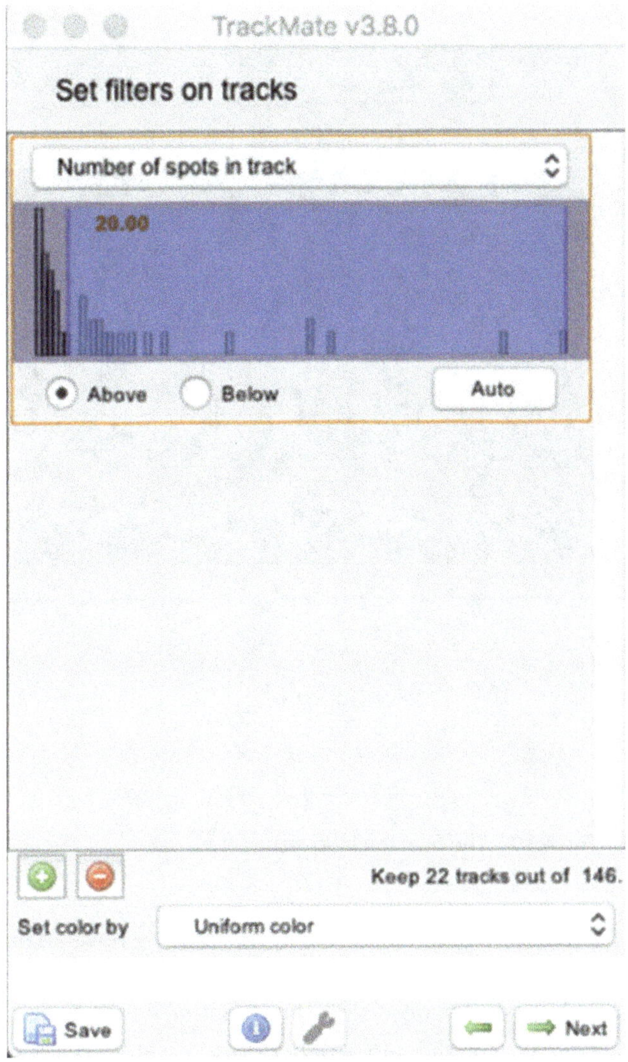

Fig. 4.9 Filtering tracks based on the number of spots they contain

Solution

Even though we do not have many tracks in our case, visual inspection is not enough. A good approach for a first comparison of tracking results is to compare tracks based on the feature values that are calculated for them. To do this, once the TrackMate GUI shows the `Display options` panel (Fig. 4.10), click on the [Analysis] button. Three tables should appear, and we just want to retain the `Track statistics` one. Duplicate it ([File > Duplicate] when the table is active) and give it a name like `Track statistics-LAP`. Then go back 3 steps, select the [Linear motion LAP tracker] and perform tracking as before. Regenerate the `Track statistics` table and compare with the previous one. You will find that we have identical tracks. The track labels will be different because they are regenerated every-time we perform tracking, but you will see they are made of the same spots.

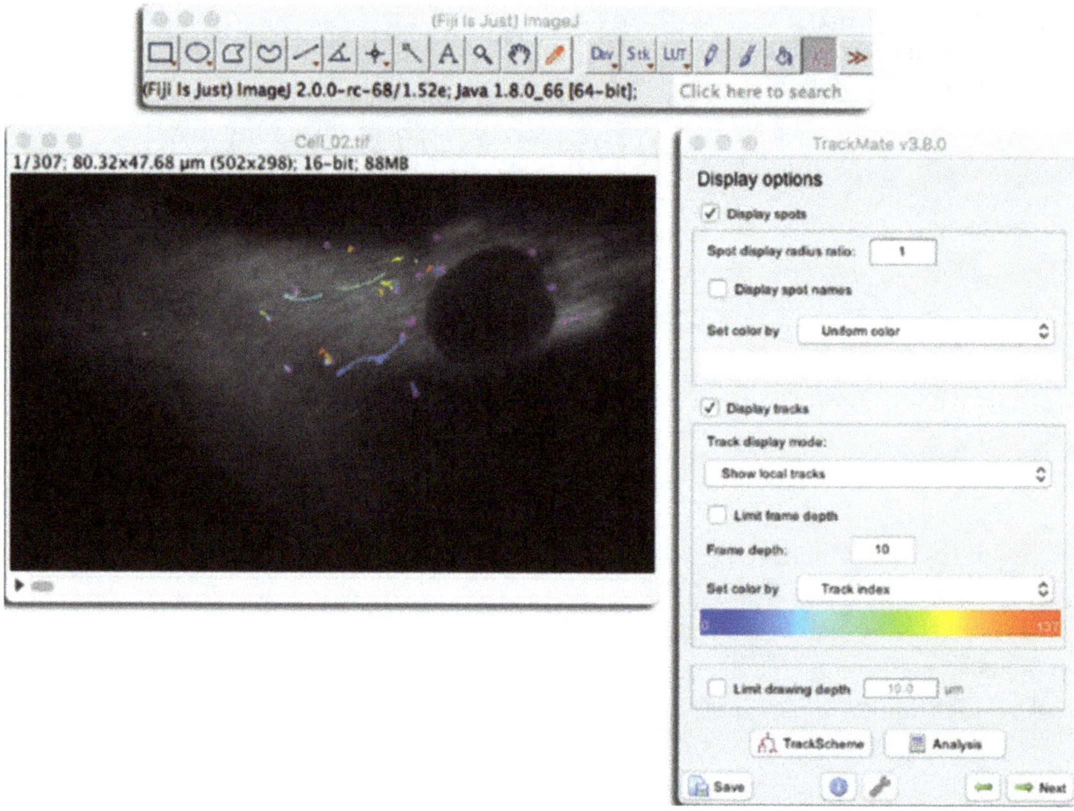

■ **Fig. 4.10** The final tracking results

This worked in our case only because we have an easy case for tracking: the spots that remain after filtering are few and well spaced. The density is so low that the motion type of the tracker does almost not matter. As shown in Chenouard et al. (2014), at high density the difference of performance among trackers is exacerbated.

4.5.6 Step 6: Export Results

We want now to export the track results in a format that can easily be re-imported in MATLAB. The XML file that is generated when you press the Save button in the GUI contains all the information to restore a tracking session: settings, parameter values, path to images, etc. It is probably not well suited to the simple track import we want to perform.

Move to the last panel of the TrackMate GUI called `Select an action`. It offers a selection of miscellaneous actions that do not fit in other panels. In the list, one action called Export tracks to XML will generate the format we want. It is a simple format derived from the one used in the ISBI single-particle tracking challenge (whose results are the subject of Chenouard et al. (2014)) and suited for tracks that do not have split nor fusion events. Execute the action and a new file called `Cell_02_Tracks.xml` should be generated. Its content looks like this:

```
<?xml version="1.0" encoding="UTF-8"?>
<Tracks nTracks="22" spaceUnits="um" frameInterval="0.5" timeUnits="s"
generationDateTime="Wed, 5 Sep 2018 14:11:51" from="TrackMate v3.8.0">
<particle nSpots="112">
<detection t="0" x="53.66873043851335" y="11.384524705860331" z="0.0" />
<detection t="1" x="53.6447201035091" y="11.417907762121915" z="0.0" />
<detection t="3" x="53.565164363562936" y="11.45756457406076" z="0.0" />
...
```


This is what we will use in MATLAB later. You can see that each track is represented by a `particle` section, containing several `detection` items, with t, x, y and z. In our case, z is always 0 since we have 2D time-series.

Exercise 4.6
Perform tracking and exports for all the other movies included in the dataset. Then move on to the next section.

4.6 Motility Analysis with Mean-Square Displacement

Tracking is almost never the last step of an analysis workflow. Tracking tools such as TrackMate produce tracks and their role stops there. But tracks are just an intermediate data structure in the workflow. Their subsequent analysis will produce the numbers upon which we will draw a scientific conclusion. Because this track analysis is specific to the scientific question to be addressed, tracking tools remain generic and seldom include specialized analysis modules. Another toolset is required for track analysis, and in this module we will focus on using MATLAB. The main reason for this choice is that there exist ready-to-use functions to import the XML files we produced in the previous section, which underlies the importance of interoperability.

There are other alternatives however. For instance, KNIME provides excellent tools to read XML, and things such as MSD of coordinates of any dimensionality can easily be computed. See for instance Hauer et al. (2017).

4.6.1 Step 1: Importing Tracks into MATLAB

Close Fiji and launch MATLAB. We want to import the tracks generated above into the MATLAB workspace. Rather than writing our own XML importer, we can use one that was made specifically for TrackMate, and that is distributed with Fiji. It is called `importTrackMateTracks` and you can find it in the `scripts` folder of Fiji:

```
tinevez@lilium:~/Development/Fiji.app/scripts$ ls
ImageJ.m                              copytoImg.m
InstallJava3D.m                       copytoImgPlus.m
IsJava3DInstalled.m                   copytoMatlab.m
Matlab3DViewerDemo_1.m                importTrackMateTracks.m
Matlab3DViewerDemo_2.m                trackmateEdges.m
Matlab3DViewerDemo_3.m                trackmateFeatureDeclarations.m
Matlab3DViewerIntroduction.m          trackmateGraph.m
Miji.m                                trackmateImageCalibration.m
Miji_Test.m                           trackmateSpots.m
bfopen.m
```

To make these scripts usable from MATLAB, open the path editor, and add the scripts folder to the path (Fig. 4.11).

This can also be achieved using `addpath('./path/to/your/Fiji.app/scripts');` in the MATLAB prompt.

Once this is done, the functions in this folder are visible and can be called from MATLAB. For instance, we can now get the help of the function we want to use in MATLAB:

```
>> help importTrackMateTracks
  |importTrackMateTracks|  Import linear tracks from TrackMate
  This function reads a XML file that contains linear tracks gener-
  ated by  TrackMate (http://fiji.sc/TrackMate). Careful: it does not
  open the XML TrackMate session file, but the track file exported in Track-
  Mate using the action 'Export tracks to XML file'. This file format
  contains less information than the whole session file, but is enough fo
  r linear tracks
  (tracks that do not branch nor fuse).
...
```

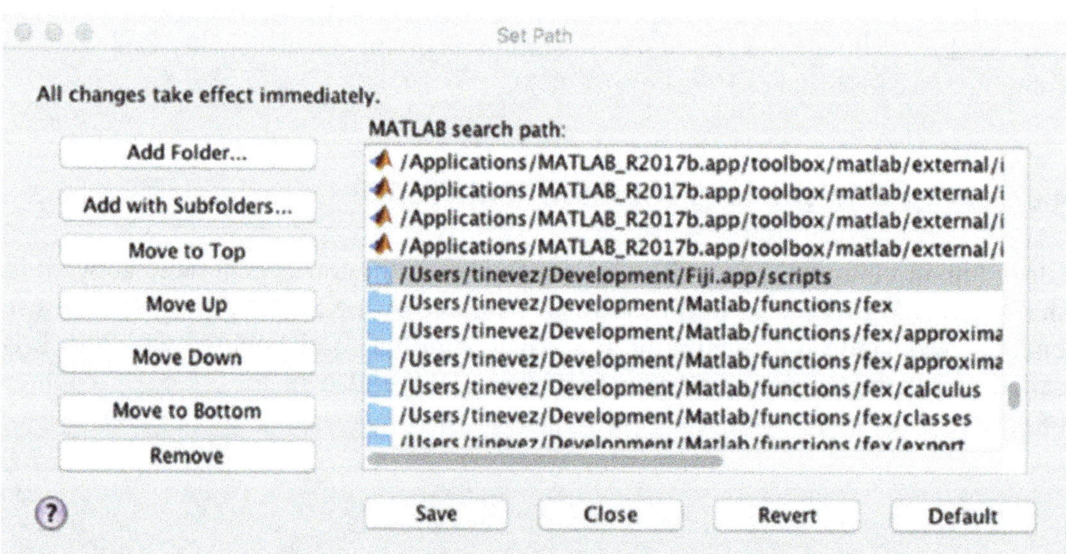

Fig. 4.11 Add the Fiji script folder to the MATLAB path

Exercise 4.7

Read the help section of the function and try to find the correct syntax to import the tracks in a desirable way. For instance, we do not need the Z coordinates, since we dealt with a 2D dataset, and we do not need the time to be scaled by a physical units.

Solution

The proper syntax is something along the lines of:

4

```
>> track_file = '/Users/tinevez/Desktop/Tracking-NEMO-movies_subset/
        NEMO-IL1/Cell_02_Tracks.xml';
>> tracks = importTrackMateTracks(track_file, true, false);
```

You need of course to specify the path to the XML file we saved in the previous section. The first flag `true` is used to specify that we do not need to import the Z coordinates, and the second flag `false` is used to specify that we want a time interval in integer units of frames.

The imported content is made of a cell list of several $N \times 3$ arrays:

```
>> tracks
tracks =
  22x1 cell array
    112x3 double
    268x3 double
    159x3 double
...
```

And each array contains 3 columns with the frame, X and Y coordinates, one line per time point of a complete track:

```
>> tracks1
    ans =
                 0   53.6687   11.3845
            1.0000   53.6447   11.4179
            3.0000   53.5652   11.4576
            4.0000   53.6317   11.3376
            5.0000   53.6501   11.3377
            6.0000   53.5482   11.4344
...
```

From it, we can plot an example trajectory:

```
>> x = tracks1(:,2);
>> y = tracks1(:,3);
>> plot(x,y, 'ko-', 'MarkerFaceColor', 'w'), axis equal, box off
```

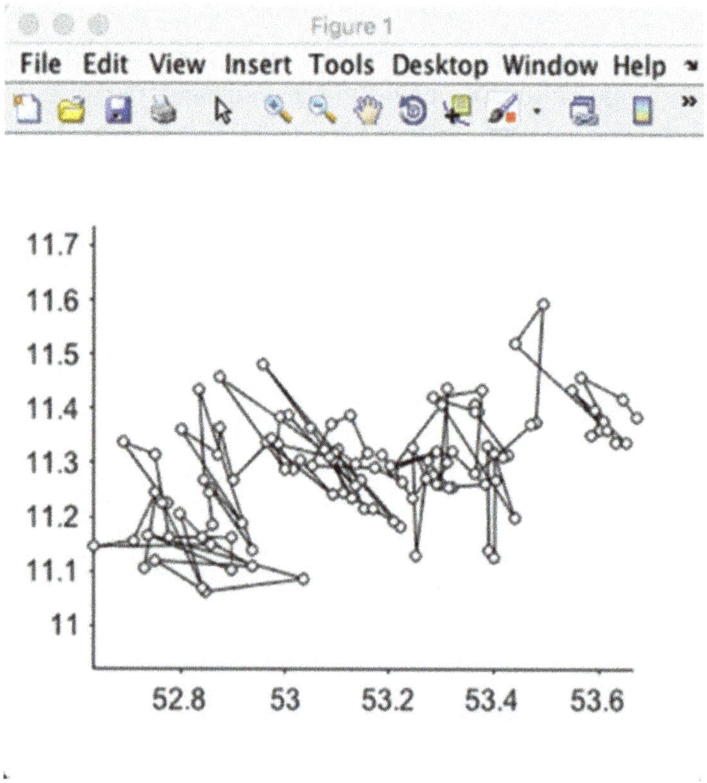

4.6.2 Step 2: Create and Add Data to the MSD Analyzer

As stated above, `@msdanalyzer` is a MATLAB class. If you do not know what is a class, you can think of it roughly as a collection of functions organized around a common and clearly defined data structure. The functions of a class are called methods and we will use this denomination in the following. If you followed the instruction of ▶ Sect. 4.3, the `@msdanalyzer` should be on the MATLAB path. You should be able to access the help for the class and the help for the constructor of the class:

```
>> help msdanalyzer
>> help msdanalyzer.msdanalyzer
```

The first instruction gives help about the class itself and the second syntax gives you help about the syntax to use when creating an analyzer. You can retrieve the list of methods defined for this class with

```
>> methods('msdanalyzer')
```

and the help for a method called `addAll` is obtained via:

```
>> help msdanalyzer.addAll
```

We need to create an analyzer first, before giving it data. This is done like this:

```
>> ma = msdanalyzer(2, 'um', 'frames')
```

Now `ma` is an empty `msdanalyzer` object, set to operate for 2D data (this is the meaning of the '2' as first argument), using µm as spatial units ('um' because MATLAB does not handle UTF8 characters very well) and frames as time units.

As stated above, this object is empty, and we have to feed it the tracks with the `addAll( )` method. Luckily for us, as you can read in the help of the `addAll` method, it expects the tracks to be formatted exactly in the shape we have. So we can run directly:

```
>> ma = ma.addAll(tracks)
    ma =
      msdanalyzer with properties:
          TOLERANCE: 12
             tracks: 22x1 cell
              n_dim: 2
        space_units: 'um'
         time_units: 'frames'
...
```

Note that for now, it just has the 22 tracks of the first movie we analyzed. We want to add the tracks coming from the other movies *in the same category*. For instance, we will later add to the same `msdanalyzer` object all the tracks coming from all the movies of the NEMO-IL1 folder. But for now, we can use some of the methods of the `msdanalyzer` to have a nice track overview:

```
>> ma.plotTracks            % Plot the tracks.
>> ma.labelPlotTracks       % Add labels to the axis.
>> set(gca, 'YDir', 'reverse')
>> set(gca, 'Color', [0.5 0.5 0.5])
>> set(gcf, 'Color', [0.5 0.5 0.5])
```

In ◘ Fig. 4.12, the results of these commands are displayed next to the TrackMate results. The track colors happen to be the same, but this is by chance. As a side note, look at the line 3 in the above snippet. When displaying images, the Y axis runs from top to bottom.

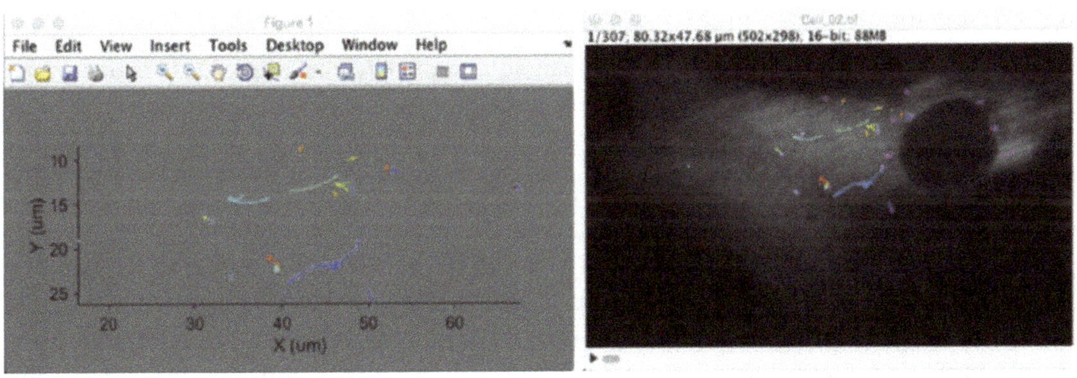

◘ **Fig. 4.12** TrackMate tracks displayed in MATLAB

But MATLAB displays data in plots where it runs in the other direction, so we had to invert it here to make the tracks look like their Fiji counterparts.

Exercise 4.8

Repeat this procedure to add all the tracks to the same `msdanalyzer` object. Be careful to use the `ma = ma.addAll( )` syntax each time.

Solution

Supposing we continue with the `msdanalyser` object we created above (and using the XML files that are already distributed with the dataset…):

```
>> tracks2   =   importTrackMateTracks('/Users/tinevez/Desktop/Track-
   ing-NEMO-movies_subset/NEMO-IL1/ Cell_03_Tracks.xml', true, false);
>> ma = ma.addAll(tracks2);
>> tracks3 = importTrackMateTracks('/Users/tinevez/Desktop/Tracking-
   NEMO-movies_subset/NEMO-IL1/ Cell_04_Tracks.xml', true, false);
>> ma = ma.addAll(tracks3);
>> ma
    ma =
      msdanalyzer with properties:
          TOLERANCE: 12
             tracks: 614x1 cell
              n_dim: 2
        space_units: 'um'
         time_units: 'frames'
...
```

4.6.3 Interlude: A Short Word About Mean-Square Displacement Analysis

Let's consider particles undergoing Brownian motion. Let's suppose that all the particles were released from a single point at $t = 0$, that r is the distance to this point, and that D is the diffusion coefficient of all these particles in the medium they diffuse in. We can find the equation for their density for instance in one of Einstein's historical papers (Einstein (1905)):

$$\rho(r,t) = \rho_0 \exp\left(-\frac{r^2}{4Dt}\right)$$

Using this formula, one can derive the mean square displacement (MSD) for such particles. After a delay τ, the mean-square displacement of the particle ensemble is:

$$\text{MSD}(\tau) = \left\langle r^2 \right\rangle = 2dD\tau \qquad (4.1)$$

We see that the plot of the MSD value as a function of time delay τ should be a straight line in the case of simple freely diffusing movement. We therefore have a way to check what is the motion type of the particles. If the MSD is a line, then it is diffusing, and the slope gives us the diffusion coefficient. If the MSD saturates and has a concave curvature, then its movement is impeded: it cannot freely diffuse away from its starting point. On the contrary, if the MSD increases faster than at linear rate, then it must be transported, because Brownian motion could not take it away that fast. See Qian et al. (1991) for a first application to biological data.

This is great, because to decide whether the erratic movement of a particle that you are observing is freely diffusive, impeded, or transported, you would only have to follow the particle for a finite amount of time. This equation can be evaluated to check what the particle movement type is. So we just need a way to evaluate it practically.

Experimentally, the MSD for a single particle is also taken as a mean. If the process is stationary (that is: the "situation", experimental conditions, etc… do not change over time) and spatially homogeneous, the ensemble average can be taken as a time average for a single trajectory, and MSD for a single particle i can be calculated as

$$r_i^2(t,\tau) = \left(r_i(t+\tau) - r_i(t)\right)^2$$

We then average over overall possible t for a given delay τ to yield $\mathrm{MSD}_i(\tau)$, and then average the resulting MSD_i over all particles. This is exactly what the `@msdanalyzer` class was built for, as we will see now.

We note that for finite trajectories, the smaller delays τ will be more represented in the average than longer delays. For instance, if a trajectory has N points in it, the delay corresponding to one frame will have $N-1$ points in the average, and the delay corresponding to N frames will only have one. This has major consequences on measurement certainty, see Michalet (2010). This is one of the reason why we insisted above on having tracks that were not too short. Additionally, one has to keep in mind that processes are rarely stationary over long period of times and anomalous diffusion (any case when the MSD is not a line) processes are families of various origins which can have more specific effects on MSD.

4.6.4 Step 3: Compute the Mean-Square Displacement

The `@msdanalyzer` automates the calculation. Using the object we prepared in step 2, calculating MSD is as simple as:

```
>> ma = ma.computeMSD
    Computing MSD of 614 tracks...    Done.
    ma =
      msdanalyzer with properties:
          TOLERANCE: 12
             tracks: 614x1 cell
              n_dim: 2
        space_units: 'um'
         time_units: 'frames'
                msd: 614x1 cell
...
```

Notice that now the `msd` field of the object has some content. However interpreting it is not trivial. The plot of the individual MSD curves look like this:

```
>> ma.plotMSD
```

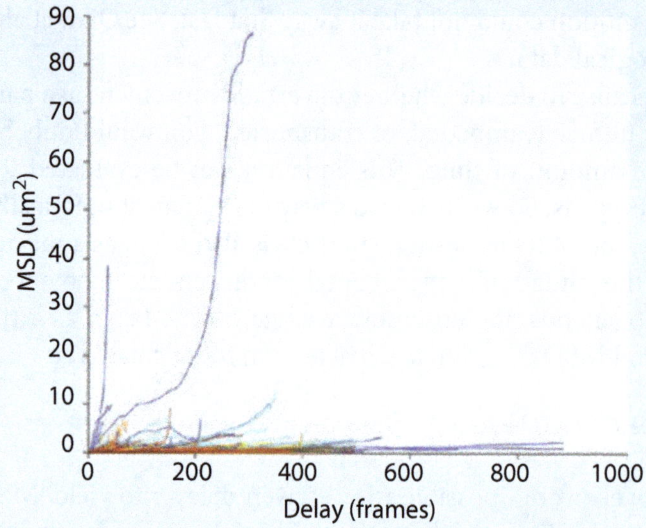

We can plot the ensemble-mean MSD, averaged over all particles:

```
>> ma.plotMeanMSD
```

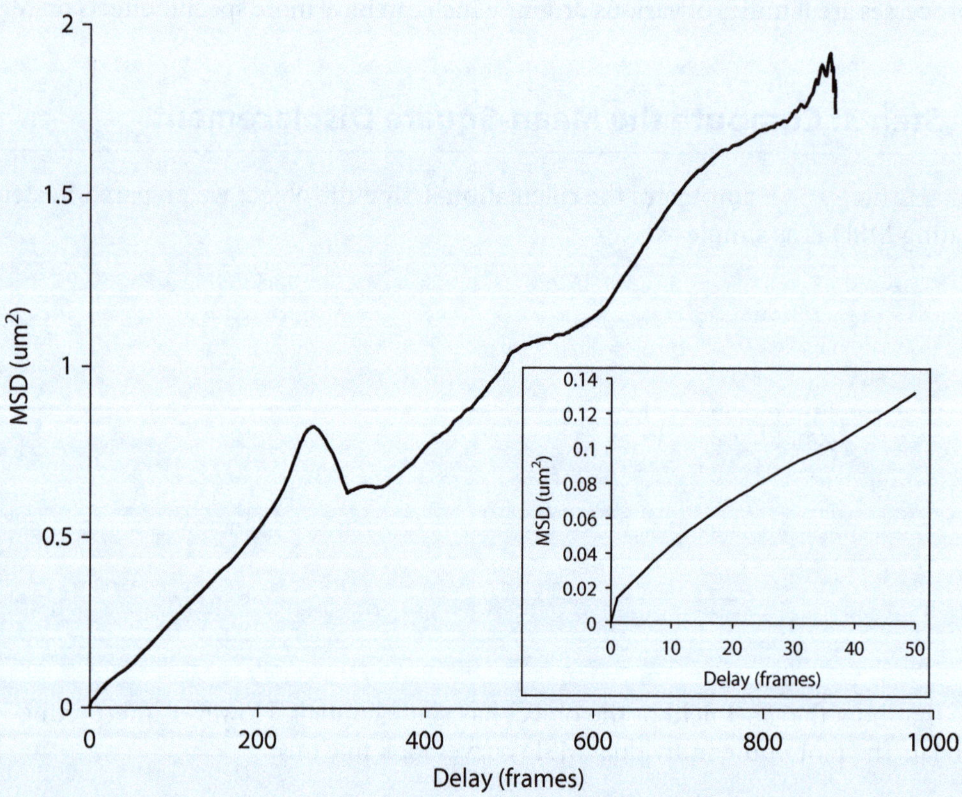

Apart from the bumps beyond a delay of 200 frames, the ensemble curve looks like a straight line. However, looking at smaller delays, we see that this curve displays a slight concavity.

4.6.5 Step 4: Log-Log Fit of the Mean-Square Displacement

4

This is not enough for us to conclude. A cell is a complex environment and each particle might have different properties that are confused in the ensemble mean plotted above. We therefore turn to another strategy.

Let's consider a single particle that diffuses freely. In that case Eq. 4.1 holds. The MSD as a function of delay τ is a straight line. If we compute the logarithm of Eq. 4.1 we get:

$$\begin{aligned} \mathrm{MSD}_{\mathtt{diff}}(\tau) &= 2dD_i\tau \\ \log\left(\mathrm{MSD}_{\mathtt{diff}}(\tau)\right) &= \log(\tau) + C \end{aligned} \tag{4.2}$$

that we can write $y = 1 \times x + C$ if y is $(\log \mathrm{MSD})$ and x is $(\log \tau)$.

Let us now consider a particle that moves with a nearly constant velocity vector. In that case, r varies linearly with τ and MSD varies with the square of τ. We then can write:

$$\begin{aligned} \mathrm{MSD}_{\mathtt{trans}}(\tau) &\propto \tau^2 \\ \log\left(\mathrm{MSD}_{\mathtt{trans}}(\tau)\right) &= 2 \times \log(\tau) + C \end{aligned} \tag{4.3}$$

or $y = 2 \times x + C$.

So in a log-log plot, the MSD curves can be approximated by straight lines of slope 1 for diffusion motion, 2 for transported motion, and less than 1 for constrained motion. We can therefore turn this into a test to determine the motion type of our dots. We will fit the log-log plot of the MSD curve by a line for each particle, and measure its slope *alpha*. The distribution of all slopes for a given condition will yield the motion type. We can also use the fitting approach to add an automated quality check. For instance, we can decide not to include slope values for fits with an R^2 lower than 0.5.

Again, there is a method that does all of this for us in the `@msdanalyzer` class:

```
% Get the description of the log-log fit function.
>> help ma.fitLogLogMSD
% Perform the fit:
>> ma = ma.fitLogLogMSD
Fitting 614 curves of log(MSD) = f(log(t)), taking only the first 25
% of each curve... Done.
% Note that now the loglogfit field of the analyzer is not empty any-
more:
>> ma.loglogfit
ans =
  struct with fields:
        alpha: [614x1 double]
        gamma: [614x1 double]
```

```
       r2fit : [614x1 double]
   alpha_lci: [614x1 double]
   alpha_uci: [614x1 double]
```

We are interested in the slope *alpha*, but first we want to remove all fits that had an R^2 value lower than 0.5. The R^2 values are stored in the `ma.loglogfit.r2fit` field.

```
% Logical indexing:
>> valid = ma.loglogfit.r2fit > 0.5;
>> fprintf('Retained %d fits over %d.\n', sum(valid), numel(valid))
Retained 461 fits over 614.
```

Now we can plot the histogram of slopes:

```
>> histogram(ma.loglogfit.alpha( valid ), 'Normalization', 'probability')
>> box off
>> xlabel('Slope of the log-log fit.')
>> ylabel('p')
>> yl = ylim;
>> line( [ 1 1 ], [ yl(1) yl(2) ], 'Color', 'k', 'LineWidth', 2)
```

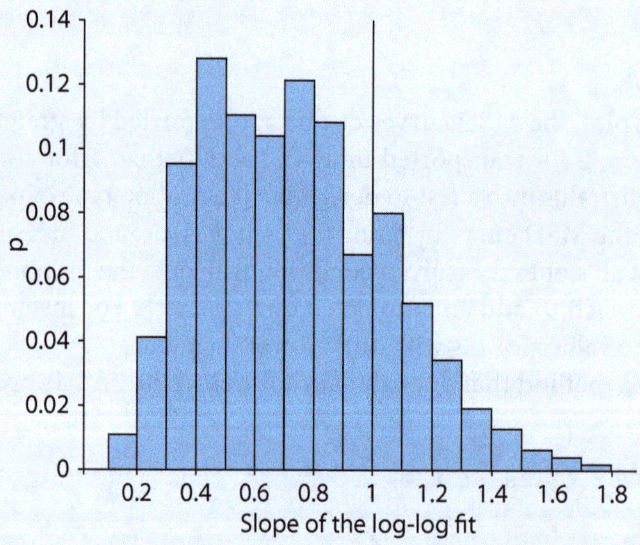

4.6.6 Step 5: Analysis of the Log-Log Fit

The histogram displayed above shows a peak around a slope of 1, and several other peaks below 1, around 0.4 and 0.8 judged from its shape. This suggests that there are mixed populations in our dataset, with some particles freely diffusing and others, the majority, probably constrained.

The population average behavior can be assessed by computing the mean of this distribution and checking whether it is significantly lower than 1 based on a t-test evaluation.

```
>> fprintf('Mean slope in the log-log fit: alpha = %.2f +/- %.2f
(N = %d).\n', ...
mean( ma.loglogfit.alpha(valid) ), std( ma.loglogfit.alpha(valid)),
sum(valid))
>> if (h)
    fprintf('The mean of the distribution IS significantly lower than
1 with P = %.2e.\n', p)
else
    fprintf('The mean of the distribution is NOT significantly
lower than 1. P = %.2f.\n', p)
end
Mean slope in the log-log fit: alpha = 0.73 +/- 0.31 (N = 461).
The mean of the distribution IS significantly lower than 1 with
P = 4.79e-57.
```


This ensemble analysis is not perfectly relevant however. The t-test we ran at the end gives a conclusion on the mean of the slope value, which is not exactly what we want to know. We know that there are likely to be a mixed population of particles with different motility. We expect for instance some non-specific particles to be freely diffusing or completely stuck to the substrate.

We may ask how many particles have a constrained motility and if they are the majority. A way to assess this at the single particle level is to check the confidence interval for the value of the slope in the fit. We state that if the confidence interval of the slope value is below 1, then the particles have a constrained motility. Again, things are made easy to us, as the confidence interval is also stored in the `@msdanalyzer` instance:

```
cibelow = ma.loglogfit.alpha_uci(valid) < 1;
ciin = ma.loglogfit.alpha_uci(valid) >= 1 & ma.loglogfit.
alpha_lci(valid) <= 1;
ciabove = ma.loglogfit.alpha_lci(valid) > 1;
fprintf('Found %3d particles over %d with a confidence interval for t
he slope value below 1.\n', ...
    sum(cibelow), numel(cibelow))
fprintf('Found %3d particles over %d with a slope of 1 inside the
confidence interval.\n', ...
    sum(ciin), numel(ciin))
fprintf('Found %3d particles over %d with a confidence interval for t
he slope value above 1.\n', ...
    sum(ciabove), numel(ciabove) )
Found 345 particles over 461 with a confidence interval for the
slope value below 1.
Found  36 particles over 461 with a slope of 1 inside the confidence
interval.
Found  80 particles over 461 with a confidence interval for the
slope value above 1.
```

This allowed us to conclude that the majority of the dots that were tracked have a constrained or sub-diffusive motility at the time-scale of their appearance. A reasonable hypothesis is that they are anchored to some static structure in the cell.

Exercise 4.9

Would the conclusion have been very different if we had been much more stringent on the R^2 value we used to filter out bad tracks? For instance, with $R^2 = 0.8$?

Solution

The distribution of alpha changes. The peak centered around 0.4 disappears, and the histogram takes the shape of a large and wide peak centered at 0.8, with a secondary, small peak around 1. The mean slope value changes accordingly, however the conclusion on the motility type is still valid.

```
>> valid = ma.loglogfit.r2fit > 0.8;
>> fprintf('Retained %d fits over %d.\n', sum(valid), numel(valid))
Retained 317 fits over 614.
>> fprintf('Mean slope in the log-log fit: alpha = %.2f +/- %.2f
(N = %d).\n', ...
mean(ma.loglogfit.alpha(valid)),         std(ma.loglogfit.alpha(valid)),
sum(valid))
>> if (h)
    fprintf('The mean of the distribution IS significantly lower than
1 with P = %.2e.\n', p)
else
    fprintf('The mean of the distribution is NOT signifi-
cantly lower than 1. P = %.2f.\n', p)
end
Mean slope in the log-log fit: alpha = 0.86 +/- 0.28 (N = 317).
The mean of the distribution IS significantly lower than 1 with
P = 2.31e-17.
```

Exercise 4.10

Redo all the analysis for the control condition. In our case, the control condition corresponds to cells that were not stimulated. The dots we observed then were permanent instead of being transient when the cells were stimulated. They probably correspond to some spurious particles.

Solution

We can re-execute the whole approach displayed above on the two movies in the `Control` folder:

```
clear all
close all
clc
tracks1 = importTrackMateTracks('/Users/tinevez/Desktop/Tracking-
NEMO-movies_subset/NEMO-Ctrl/ Cell_01_Tracks.xml', true, false);
tracks2 = importTrackMateTracks('/Users/tinevez/Desktop/Tracking-
NEMO-movies_subset/NEMO-Ctrl/ Cell_02_Tracks.xml', true, false);
ma = msdanalyzer(2, 'um', 'frames');
```

```
ma = ma.addAll(tracks1);
ma = ma.addAll(tracks2);
ma = ma.computeMSD;
ma = ma.fitLogLogMSD;
valid = ma.loglogfit.r2fit > 0.8;
fprintf('Retained %d fits over %d.\n', sum(valid), numel(valid))
fprintf('Mean slope in the log-log fit: alpha = %.2f +/- %.2f (N = %d).
\n', ...
  mean(ma.loglogfit.alpha(valid)), std(ma.loglogfit.alpha(valid)), sum
(valid))
[h, p] = ttest( ma.loglogfit.alpha(valid), 1, 'tail', 'left');
if (h)
     fprintf('The mean of the distribution IS signifi-
cantly lower than 1 with P = %.2e.\n', p)
else
     fprintf('The mean of the distribution is NOT signifi-
cantly lower than 1. P = %.2f.\n', p)
end
```


And the output is:

```
Computing MSD of 19 tracks...   Done.
Fitting 19 curves of log(MSD) = f(log(t)), taking only the first 25%
of each curve... Done.
Retained 13 fits over 19.
Mean slope in the log-log fit: alpha = 0.87 +/- 0.24 (N = 13).
The mean of the distribution IS significantly lower than 1 with
P = 3.41e-02.
```

The majority of non-specific particles appears to also be stuck. So what is the difference with the IL1-stimulation condition? In this case, the conclusion on motility is the same but it does not apply to the same particles. The control condition movies are made of the few cells we could find that had fluorescent dots that were visible without stimulation. What matters is that there are few of them and that their number is not enough to change the scientific conclusion on the dots that appear transiently upon stimulation, regardless of their motility.

4.7 Results and Conclusion

This module is one part of the work that helped us conclude on the NEMO dot motility. The MSD analysis indicated that the dots made of NEMO-eGFP proteins that appear upon stimulation by IL1 are anchored to some static structure of the cell during the little time they are visible.

We then turned to investigate what this static structure could be. So we repeated the analysis you just did on cells for which we depolymerized actin filaments and microtubules. The conclusion did not change. There was still the same proportion of NEMO dots with the same constrained motility type.

We then led other investigations, relying on biochemistry and confocal imaging, and concluded that NEMO dots are anchored at the cell membrane. The membrane is fluid and the anchor point might be diffusing, but we do not see this behavior on the time-scale of the live-imaged NEMO dots. The whole story and more can be found in the original paper Tarantino et al. (2014).

Take Home Message

We hope this module serves as an example and shows that biophysics and image analysis can provide new approaches to a scientific question that would otherwise solely rely on biochemistry. The original paper contains quite some heavy biochemistry studies, but their results are reinforced by the orthogonal approach presented here.

We relied on mean-squared-displacement analysis to reach a conclusion on the motility type. This is the historical method and the first to have been applied on biological data Qian et al. (1991). Its main drawback is that the tracks need to be long and the detections accurate to have a decent accuracy on the quantities MSD analysis yields. Good, accurate tracks are especially difficult to obtain in many life-science cases, so several research labs have been working on developing new methods improving on MSD. We can cite for instance the work of Hansen et al. (2018) based on analyzing step distributions, or Briane et al. (2018) that relies on a statistical approach.

Acknowledgements We are very grateful to Emmanuel Laplantine and Nadine Tarantino, our fellow authors on the paper, that agreed to make the raw data publicly available. We thank Jan Eglinger (Friedrich Miescher Institute for Biomedical Research, Basel) for reviewing this chapter.

Bibliography

Briane V, Kervrann C, Vimond M (2018) Statistical analysis of particle trajectories in living cells. Phys Rev E 97:062121 https://doi.org/10.1103/PhysRevE.97.062121. https://link.aps.org/doi/10.1103/PhysRevE.97.062121

Chenouard N, Smal I, de Chaumont F, Maška M, Sbalzarini IF, Gong Y, Cardinale J, Carthel C, Coraluppi S, Winter M, Cohen AR, Godinez WJ, Rohr K, Kalaidzidis Y, Liang L, Duncan J, Shen H, Xu Y, Magnusson KE, Jalden J, Blau HM, Paul-Gilloteaux P, Roudot P, Kervrann C, Waharte F, Tinevez JY, Shorte SL, Willemse J, Celler K, van Wezel GP, Dan HW, Tsai YS, Ortiz de Solorzano C, Olivo-Marin JC, Meijering E (2014) Objective comparison of particle tracking methods. Nat Methods 11(3):281–289

Einstein A (1905) Investigations on the theory of the brownian movement. Ann der Physik. http://www.physik.fu-berlin.de/~kleinert/files/eins_brownian.pdf

Hansen AS, Woringer M, Grimm JB, Lavis LD, Tjian R, Darzacq X (2018) Robust model-based analysis of single-particle tracking experiments with spot-on. eLife 7:e33125. ISSN: 2050-084X. https://doi.org/10.7554/eLife.33125

Hauer MH, Seeber A, Singh V, Thierry R, Sack R, Amitai A, Kryzhanovska M, Eglinger J, Holcman D, Owen-Hughes T, Gasser SM (2017) Histone degradation in response to DNA damage enhances chromatin dynamics and recombination rates. Nat Struct Mol Biol 24(2):99–107

Jaqaman K, Loerke D, Mettlen M, Kuwata H, Grinstein S, Schmid SL, Danuser G (2008) Robust single-particle tracking in live-cell time-lapse sequences. Nat Methods 5(8):695–702

Kalman RE (1960) A new approach to linear filtering and prediction problems. Trans ASME J Basic Eng 82(Series D):35–45

Michalet X (2010) Mean square displacement analysis of single-particle trajectories with localization error: Brownian motion in an isotropic medium. Phys Rev E Stat Nonlin Soft Matter Phys 82(4 Pt 1):041914

Qian H, Sheetz MP, Elson EL (1991) Single particle tracking. analysis of diffusion and flow in two-dimensional systems. Biophys J 60(4):910–921. ISSN: 0006-3495. https://doi.org/10.1016/S0006-3495(91)82125-7. http://www.sciencedirect.com/science/article/pii/S0006349591821257

Tarantino N, Tinevez J-Y, Crowell EF, Boisson B, Henriques R, Mhlanga M, Agou F, Israël A, Laplantine E (2014) TNF and IL-1 exhibit distinct ubiquitin requirements for inducing NEMO–IKK supramolecular structures. J Cell Biol 204(2):231–245. ISSN: 0021-9525. https://doi.org/10.1083/jcb.201307172. http://jcb.rupress.org/content/204/2/231

Tinevez J-Y, Dragavon J, Baba-Aissa L, Roux P, Perret E, Canivet A, Galy V, Shorte S (2012) A quantitative method for measuring phototoxicity of a live cell imaging microscope, chap 15. In: Michael Conn P (ed) Imaging and spectroscopic analysis of living cells. Methods in enzymology, vol 506, pp 291–309. Academic. https://doi.org/10.1016/B978-0-12-391856-7.00039-1. http://www.sciencedirect.com/science/article/pii/B9780123918567000391

Tinevez J-Y, Perry N, Schindelin J, Hoopes GM, Reynolds GD, Laplantine E, Bednarek SY, Shorte SL, Eliceiri KW (2017) TrackMate: an open and extensible platform for single-particle tracking. Methods 115: 80–90. ISSN: 1046-2023. https://doi.org/10.1016/j.ymeth.2016.09.016. http://www.sciencedirect.com/science/article/pii/S1046202316303346 Image Processing for Biologists

Introduction to MATLAB

Image Analysis and Brownian Motion

Simon F. Nørrelykke

K. Miura, N. Sladoje (eds.), *Bioimage Data Analysis Workflows*, Learning Materials in Biosciences,
https://doi.org/10.1007/978-3-030-22386-1_5

What You Learn from This Chapter

You will be introduced to some of the powerful and flexible image-analysis methods native to MATLAB. You will also learn to use MATLAB to simulate a time-series of Brownian motion (diffusion), to analyse time-series data, and to plot and export the results as pretty figures ready for publication. If this is the first time you code, except from writing Macros in ImageJ, then this will also serve as a crash course in programming for you.

5.1 Tools

We shall be using the commercial software package MATLAB as well as some of its problem specific toolboxes, of which there are currently more than 30.

5.1.1 MATLAB

Don't panic! MATLAB is easy to learn and easy to use. But you do still have to learn it. MATLAB is short for *matrix laboratory*, hinting at why MATLAB is so popular in the imaging community—remember that an image is just a matrix of numbers. MATLAB is commercial software for numerical, as opposed to symbolic, computing. This material was developed and tested using versions R2015b, R2016a, R2017a, and R2018a of MATLAB.

5.1.2 Image Processing Toolbox

Absolutely required if you want to use MATLAB for image analysis.

5.1.3 Statistics and Machine Learning Toolbox, Curve Fitting Toolbox

Somewhat necessary for data-analysis, though we can get quite far with the core functionalities alone.

5.2 Getting Started with MATLAB

That is what we are doing here! However, if you have to leave now and still want an interactive first experience: Head over here, sign up, and take a free, two hour, interactive tutorial that runs in your web-browser and does not require a MATLAB license (they also have paid in-depth courses).

5.2.1 Baby Steps

Start MATLAB and lets get going! When first starting, you should see something similar to ◻ Fig. 5.1

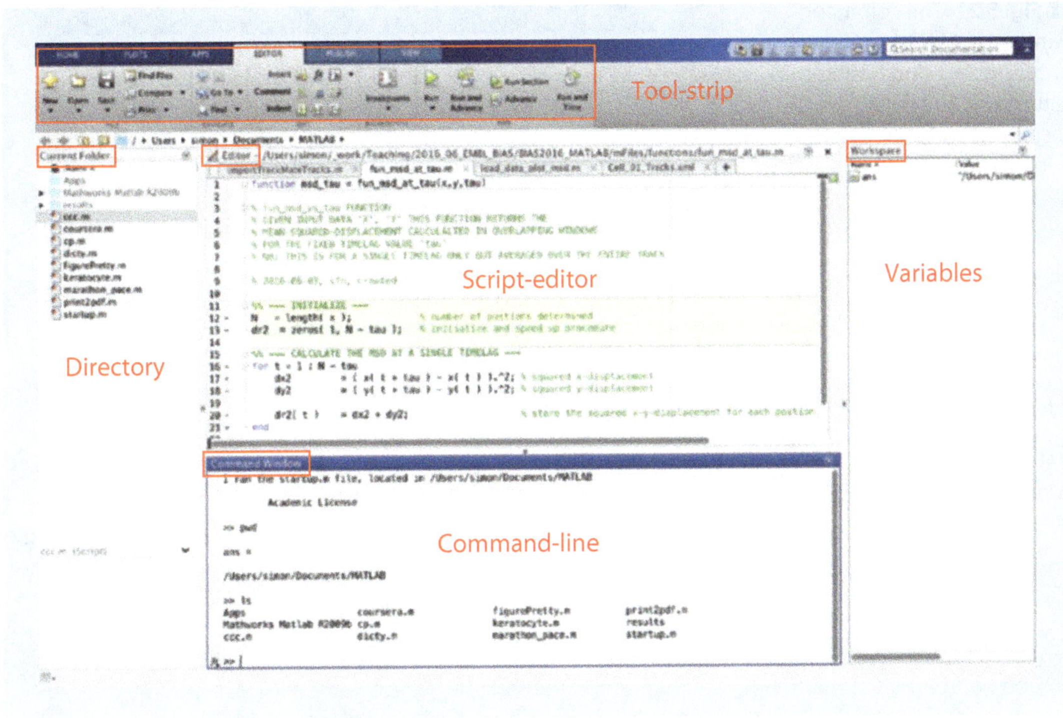

Fig. 5.1 The full MATLAB window with default layout of the windows. Some preset layouts are accessible in the tool-strip, under the `HOME` tab, in the `Layout` pull-down menu. Double-click on the top-bar of any sub-window to maximize it, double-click again to revert

First we are just going to get familiar with the command line interface. To reduce clutter, double-click on the bar (grey or blue) saying `Command Window`. This will, reversibly, maximize that window.

Now, let us add two numbers by typing `5+7`, followed by return. The result should look like in Fig. 5.2

Next, let us define two variables `a` and `b` and add them to define a third variable `c`

```
>> a=5

a =

     5

>> b=7

b =

     7

>> c=a+b

c =

    12
```

Fig. 5.2 The command window in MATLAB after entering 5+7 and hitting the return key. The result, 12, is displayed and stored in the variable **ans**

This time, we notice that the result of our operation are no longer stored in the variable **ans** but in the variable with the name we gave it, i.e., a, b, and c.

Finally, let us change one of the variables and see how the other two change in response to this.

```
>> a=10

a =

   10

>> c

c =

   12

>> c=a+b

c =

   17
```

Here, you should notice that the value of c does not change until we have evaluated it again—computers are fast, but they cannot not read our minds (most of the time), so we have to tell them *exactly* what we want them to do.

Ok, that might have been somewhat underwhelming. Let us move on to something slightly more interesting and that you can probably not so easily do on your phone.

5.2.2 Plot Something

Here are the steps we will take:

1. Create a vector x of numbers
2. Create a function y of those numbers, e.g. the cosine or similar
3. Plot y against x
4. Label the axes and give the plot a title
5. Save the figure as a pdf file

First we define the peak-amplitude (half of the peak-to-peak amplitude)

```
>> A = 10

A =

    10
```

Then we define a number of discrete time-points


```
>> x = 0 : 0.01 : 5*pi;
```

Notice how the input we gave first, the A, was again confirmed by printing (echoing) the variable name and its value to the screen. To suppress this, simply end the input with a semicolon, like we just did when defining x. The variable x is a vector of numbers, or time-points, between 0 and 5π in steps of 0.01. Next, we calculate a function $y(x)$ at each value of x

```
>> y = A * cos( x );
```

Finally, we plot y versus x[1]

```
>> figure;plot(x,y)
```

To make the figure a bit more interesting we now add one more plot as well as legend, labels, and a title. The result is shown in Fig. 5.3.

```
>> y2 = y .* x;
>> hold on
>> plot(x, y2, '--r')
>> legend('cos(x)', 'x * cos(x)')
>> xlabel('Time (AU)')
>> ylabel('Position (AU)')
>> title('Plots of various sinusoidal functions')
```

Here, **hold** on ensures that the plots already in the figure are "held", i.e., not erased, when the next function is plotted in the same figure window. We specify the use of a dashed red line, for the new plot, by the `'--r'` argument in the **plot** function. You

1 By now, you have probably noticed that some words are **typeset like this**. Those are words that MATLAB recognize as commands (excluding commands that are specific to toolboxes).

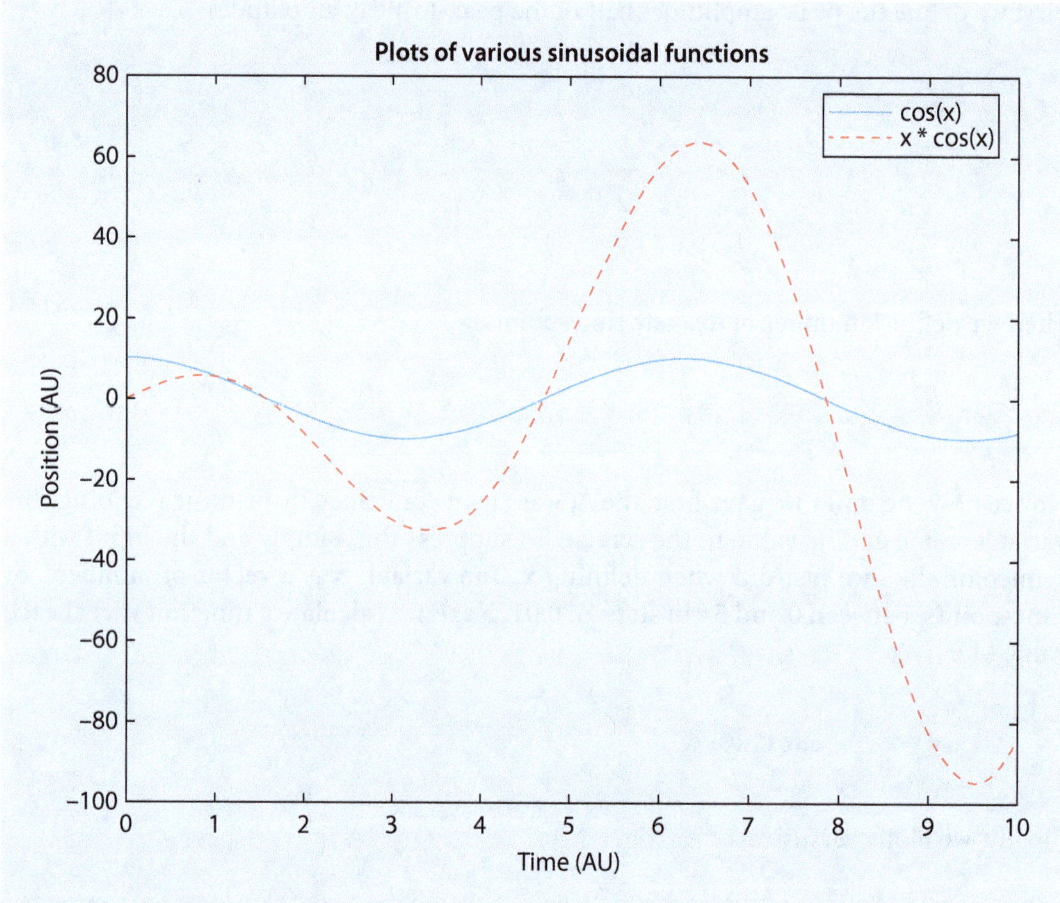

Fig. 5.3 Two sinusoidal plots with legend, axes labels, and title

will also have noticed that we multiplied using .∗ and not just ∗—this is known as element-wise multiplication, as opposed to matrix or vector multiplication (more on that in a little while).

After having created a figure and adjusted it to your liking, you may want to export it for use in a paper or presentation. This can be done either via the pull-down menus, if you only need to do it once, or via the command line if it is a recurrent job:

```
>> print('-dpdf', '/Users/simon/Desktop/cosineFigure.pdf')
```

Here, the first argument, `-dpdf'`, specifies the output file format; whereas the second argument specifies where (`/Users/simon/Desktop/`) the output file should be saved and with what name (`cosineFigure.pdf`). The **`print`** function is not confined to the pdf format but can also export to png, tiff, jpeg, etc. On a Windows machine, the path to the desktop is something like `c:Users$username)Desktop`, though it will depend on the version of Windows you run.

5.2.3 Make it Pretty

We have a large degree of control over how things are rendered in MATLAB. It is possible to set the typeface, font, colors, line-thickness, plot symbols, etc. Don't overdo this! The main objective is to communicate your message, and that message is rarely "look how many colors I have"—if you only have two graphs in the same figure, gray-scale will likely suffice. Strive for clarity!

5.2.4 Getting Help

At this point you might want to know how to get help for a specific command. That is easy, simply type **help** and then the name of the command you need help on. Example, for the **xlabel** command we just used:

```
>> help xlabel
  xlabel X-axis label.
   xlabel('text') adds text beside the X-axis on the current axis.

     xlabel('text',    'Property1',PropertyValue1,    'Property2',
     PropertyValue2 ,...)
     sets the values of the specified properties of the xlabel.

     xlabel(AX,...) adds the xlabel to the specified axes.

     H = xlabel(...) returns the handle to the text object
     used as the label.

     See also ylabel, zlabel, title, text.

     Reference page for xlabel
```

If you click the link on the last line it will open a separate window with more information and graphical illustrations. Alternatively, simply go directly to that page this way

```
>> doc xlabel
```

Expect to spend substantial time reading once you start using more of the options available. MATLAB is a rich language and most functions have many properties that you can tune to your needs, when these differ from the default.

5.3 Automating It: Creating Your Own Programs

The command-line is a wonderful place to quickly try out new ideas—just type it in and hit return. Once these ideas become more complex we need to somehow record them in one place so that we can repeat them later without having to type everything again. You know what we are getting to: The creation of computer programs.

In the simplest of cases we can take a series of commands, that were executed in the command line, and save them to a file. We could then, at a later stage, open that file and copy these lines into the command line, one after the other, and press return. This is actually a pretty accurate description of what takes place when MATLAB runs a script: It goes through each line of the script and tries to execute it, one after the other, starting at the top of the file.

5.3.1 Create, Save, and Run Scripts

You can use any editor you want for writing down your collection of MATLAB statements. For ease of use, proximity, uniformity, and because it comes with many powerful extra features, we shall use the editor that comes with MATLAB. It will look something like in Fig. 5.4 for a properly typeset and documented program. You will recognize most of the

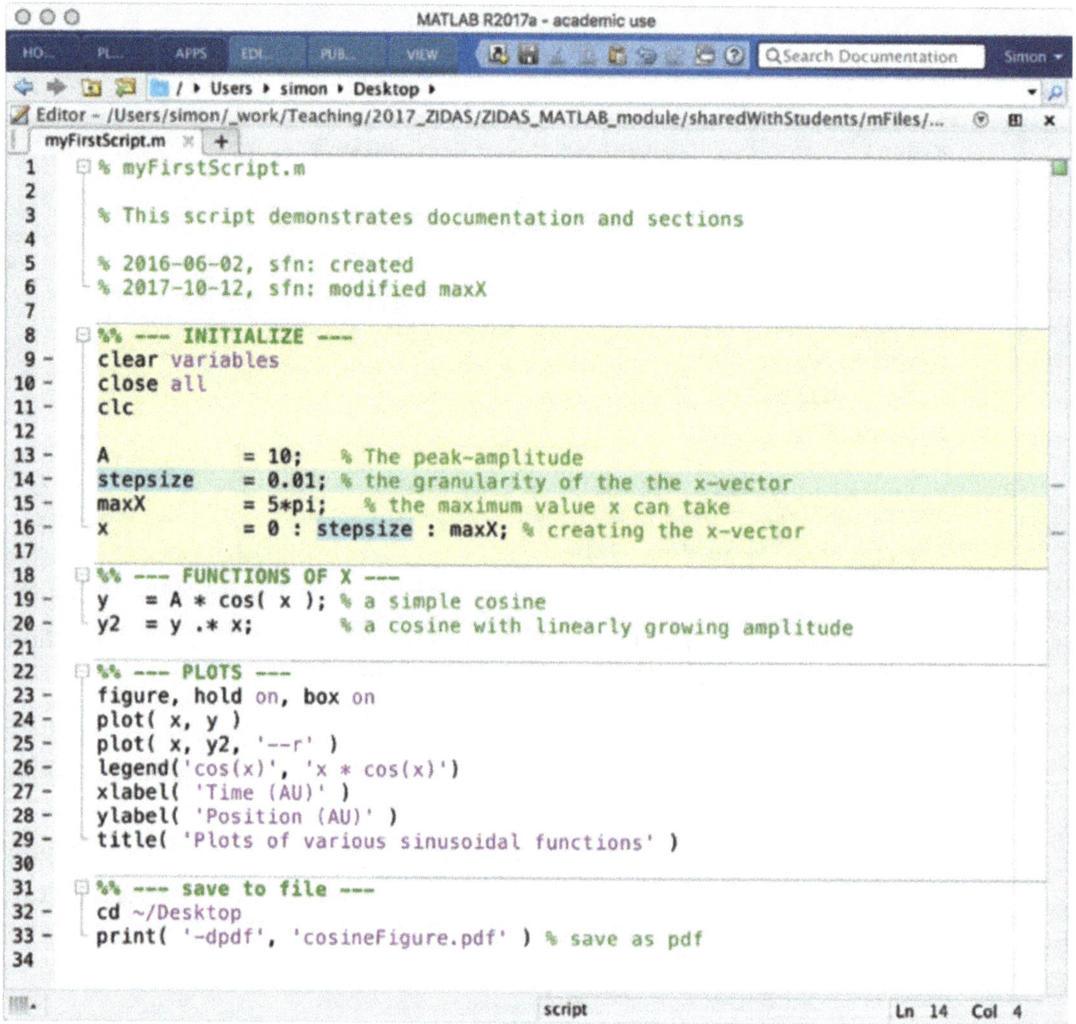

Fig. 5.4 The editor window. The script is structured for easy human interpretation with clear blocks of code and sufficient documentation. Starting from the percent sign all text to the right of it is "outcommented" and appears green, i.e., MATLAB does not try to execute it. A double percent-sign followed by space indicates the beginning of a code-block that can be folded (`command-.`), un-folded (`shift-command-.`) and executed (`command-enter`) independently. The currently active code-block is yellow. The line with the cursor in it is pale-green. Notice the little green square in the upper right corner, indicating that MATLAB is happy with the script and has no errors, warnings, or suggestions

commands from when we plotted the sinusoidsal functions earlier. But now we have also added some text to explain what we are doing.

A script like the one in Fig. 5.4 can be run in several ways: (1) You can click on the big green triangle called "run" in Editor tab; (2) Your can hit `F5` when your cursor is in the editor window; or (3) You can call the script by name from the command line, in this case simply type `myFirstScript` and hit return. The first two options will first save any changes to your script, then execute it. The third option will execute the version that is saved to disk when you call it. If a script has unsaved changes an asterisk appears next to its name in the tab.

5

When you save a script, please give it a meaningful name—"untitled.m" or "script5.m" are not good names even if you intend to never use them again (if it is temporary call it "scratch5.m" or "deleteMe5.m" so that if you forget to delete it now, you will not be in doubt when you rediscover it weeks from now). Make it descriptive and use underscores or camel-back notation as in "my_first_script.m" or "myFirstScript.m". The same goes for variable names.

5.3.2 Code Folding and Block-Wise Execution

As you will have noticed, in the screenshot of the editor, the lines of codes are split into paragraphs separated by lines that start with two percent signs and a blank space. All the code between two such lines is called a code-block. These code-blocks can be folded by clicking on the little square with a minus in it on the left (or use the keyboard shortcut `command-.`, to unfold do `shift-command-.`). This is very useful when your code grows.

You can quickly navigate between code-blocks with `command-arrow-up/down` and once your cursor is in a code-block you are interested in you can execute that entire block with `command-enter`. Alternatively, you can select (double-click or click-drag) code and execute it with `shift-F7`. For all of these actions you will see the code appearing and attempting to execute in the command window.

A list of keyboard shortcuts as well as settings for code-folding can be found in the preference settings (can you find the button?), via the `command-,` shortcut, as always, on a mac. What is it on a PC?

5.3.3 Scripts, Programs, Functions: Nomenclature

Is it a script or a program? It depends! Traditionally, only compiled languages like C, C++, Fortran, and Java are referred to as programming languages and you write programs. Languages such as JavaScript and Perl, that are not compiled, were called scripting languages and you write scripts. Then there is Python, sitting somewhere in between. MATLAB also is in between, here is what MathWorks have to say about it;

> When you have a sequence of commands to perform repeatedly or that you want to save for future reference, store them in a program file. The simplest type of MATLAB program is a script, which contains a set of commands exactly as you would type them at the command line.

Ok, so when we save our creations to an `m`-file (a file with extension .m) we call it a program file (it is a file and it is being used by the program MATLAB). But the thing we saved could be either a script or a function, or perhaps a new class definition. We shall use the word "program" to refer to both scripts and functions, basically whatever we have in the editor, but may occasionally specify which of the two we have in mind if it makes things clearer.

5.4 Working with Images

Because MATLAB was designed to work with matrices of numbers it is particularly well-suited to operate on images. Recently, Mathworks have also made efforts to become more user-friendly. Let's demonstrate (Figs. 5.5 and 5.6):

1. Save an image to your desktop, e.g. "Blobs (25K)" from ImageJ as "blobs.tif" (also provided with material)

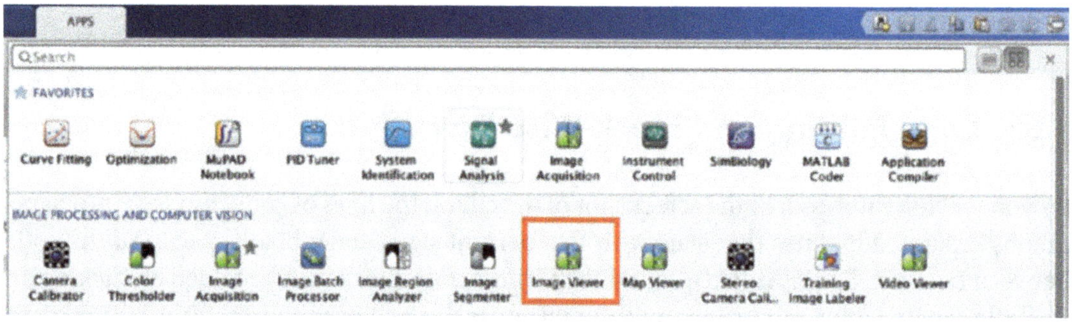

Fig. 5.5 Access to various apps in the tool-strip of MATLAB. The apps accessible will depend on the tool-boxes you have installed

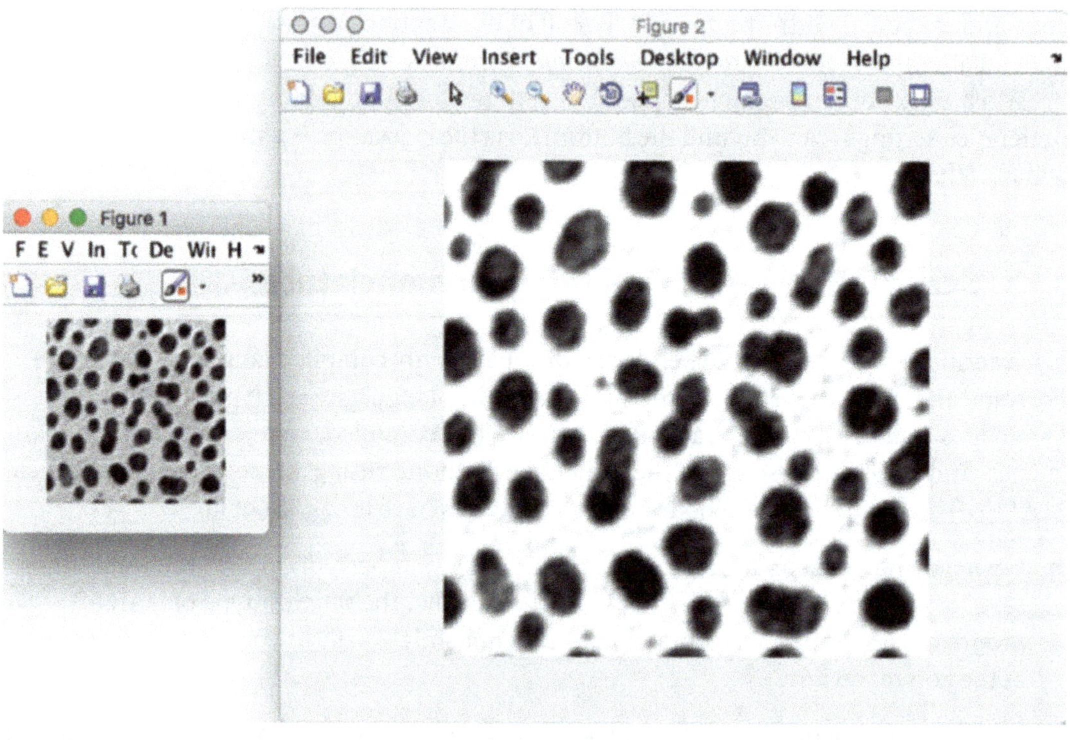

Fig. 5.6 The "blobs" from ImageJ displayed without (left) and with (right) scaling of intensity and size

2. Open the MATLAB app `Image Viewer` either from the tool-strip or by typing `imtool`
3. From the `Image Viewer` go to `File > Open ...` and select an image
4. Adjust the contrast, inspect the pixels, measure a distance, etc, using the tool-strip shortcuts

5.4.1 Reading and Displaying an Image

This, however, is not much different from what we can do in ImageJ. The real difference comes when we start working from the command-line and making scripts—while this is also possible in ImageJ, it is a lot easier in MATLAB. Assuming you have an image named "blobs.tif" on your desktop, try this

```
>> cd /Users/simon/Desktop
>> myBlobs = imread('blobs.tif');
>> figure(1); clf
>> imshow(myBlobs)
>> figure(2); clf
>> imshow(myBlobs, 'displayrange', [10 200], ...
 'initialmagnification', 'fit')
```

Here is what we just did: (1) We navigated to the directory holding our image; (2) Read the image into the variable `myBlobs` using the `imread` command; (3) Selected figure number 1 (or created it if it didn't exist yet) and cleared it; (4) Displayed the content of our variable `myBlobs` in figure 1; (5) Selected, or created, figure number 2 and cleared it; (6) Again displayed the content of `myBlobs` but now with the displayed gray-scale confined (especially relevant for 16bit images that otherwise appear black), and the displayed image fitted to the size of the window.

5.4.2 Extracting Meta-Data from an Image

Because we are becoming serious image-analysts we also take a look at the meta-data that came with the image.

```
>> blobInfo = imfinfo('blobs.tif');
>> whos blobInfo
  Name          Size            Bytes  Class     Attributes

  blobInfo      1x1              5908  struct

>> blobInfo

blobInfo =

```

```
                 Filename: '/Users/simon/Desktop/blobs.tif'
              FileModDate: '05-Jun-2016 09:45:04'
                 FileSize: 65172
                   Format: 'tif'
            FormatVersion: []
                    Width: 256
                   Height: 254
                 BitDepth: 8
                ColorType: 'grayscale'
          FormatSignature: [77 77 0 42]
                ByteOrder: 'big-endian'
           NewSubFileType: 0
            BitsPerSample: 8
              Compression: 'Uncompressed'
PhotometricInterpretation: 'WhiteIsZero'
             StripOffsets: 148
          SamplesPerPixel: 1
             RowsPerStrip: 254
          StripByteCounts: 65024
              XResolution: []
              YResolution: []
           ResolutionUnit: 'Inch'
                 Colormap: []
      PlanarConfiguration: 'Chunky'
                TileWidth: []
               TileLength: []
              TileOffsets: []
           TileByteCounts: []
              Orientation: 1
                FillOrder: 1
         GrayResponseUnit: 0.0100
           MaxSampleValue: 255
           MinSampleValue: 0
             Thresholding: 1
                   Offset: 8
         ImageDescription: 'ImageJ=1.50b...'
```

After your experience with ImageJ you should have no problems understanding this information. What is new here, is that the variable `blobInfo` that we just created is of the type `struct`. Elements in such variables can be addressed by name, like this:

```
>> blobInfo.Offset

ans =

     8

>> blobInfo.Filename

ans =

/Users/simon/Desktop/blobs.tif
```

If you want to add a field, or modify one, it is done like this:

```
>> blobInfo.TodaysWeather =  'rainy, sunny, whatever'

blobInfo =

    TodaysWeather:  'rainy, sunny, whatever'
```


Note, that we are modifying the content of the variable inside of MATLAB—the information in the "blobs.tif" file sitting on your hard-drive was not changed. If you want to save the changes you have made to an image (not including the metadata) you need the command `imwrite`. If you want to also save the metadata, and generally want more detailed control of your tif-image, you need the `Tiff` command.

When addressing an element by name, you can reduce typing by hitting the TAB-key after entering `blobInfo.`—this will display all the field-names in the structure.

It is important to realize that `imread` will behave different for different image formats. For example, the tiff format used here supports the reading of specific images from a stack via the `'index'` input argument (illustrated below) and extraction of pixel regions via the `'pixelregion'` input argument. The latter is very useful when images are large or many as it can speed up processing not having to read the entire image image into memory. On the other hand, jpeg2000 supports `'pixelregion'` and `'reductionlevel'`, but not `'index'`.

5.4.3 Reading and Displaying an Image-Stack

Taking one step up in complexity we will now work with a stack of tiff-files instead. These are the steps we will go through

1. Open "MRI Stack (528K)" in ImageJ (File > Open Samples)—or use the copy provided
2. Save the stack to your desktop, or some other place where you can find it (File > Save)
3. Load a single image from the stack into a two-dimensional variable
4. Load multiple images from the stack into a three-dimensional variable
5. Browse through the stack using the `implay` command
6. Create a montage of all the images using the `montage` command

After performing the first two steps in ImageJ, we switch to MATLAB to load a single image-plane (we will work in the editor, use `Live Script` if you feel like it) and display it (see. result in Fig. 5.7):

```
%% --- INITIALIZE ---
clear variables  % clear all variables in the workspace
close all  % close all figure windows
clc  % clear the command window
cd('~/Desktop')  % change directory to desktop
```

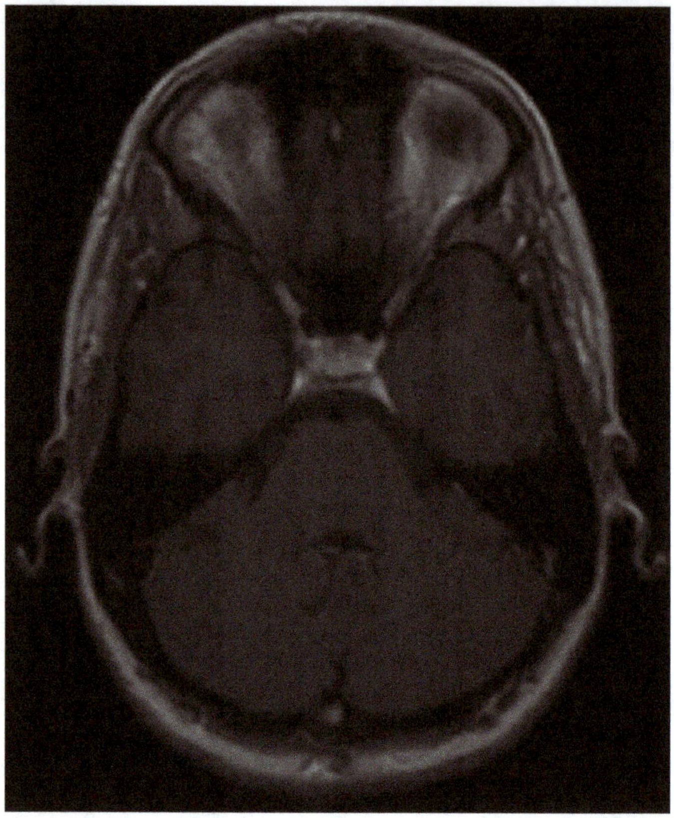

Fig. 5.7 Slice number 7 from `mri-stack.tif`

```
%% --- load single image and display it ---
mriImage = imread('mri-stack.tif',  'index', 7);
imshow(mriImage)
```

To build a stack in MATLAB we need the extra argument `'index'` to specify which single image to read and where in the stack to write it, here we chose image number 7:

```
mriStack( : , : , 7 ) = imread('mri-stack.tif', 'index',7);
```

Next, we load the entire mri-stack one image at a time. This is done by writing into the three-dimensional array (data-cube) `mriStack` using a **`for`**-loop (this concept should already be familiar to you from the ImageJ macro sections). We use the colon-notation to let MATLAB know that it should assign as many rows and columns as necessary to fit the images. We also take advantage of already knowing that there are 27 images.

```
for imageNumber = 1 : 27
        mriStack(: , : , imageNumber) = imread('mri-stack.tif',
'index', imageNumber);
end
```

We can use the **whos** command to inspect our variables and the `implay` command to loop through the stack (command line):

```
>> whos
  Name              Size              Bytes    Class     Attributes

  imageNumber       1x1                   8    double
  mriImage        226x186             42036    uint8
  mriStack        226x186x27        1134972    uint8
>> implay(mriStack)
```


Finally, we want to create a montage. This requires one additional step because we are working on 3-dimensional single-channel data as opposed to 4-dimensional RGB images (the fourth dimension is color)—the `montage` command assumes/requires 4D data (that is just how it is):

```
mriStack2    = reshape(mriStack, [226 186 1 27]);
map          = colormap('copper');  % or: bone, summer, hot
montage(mriStack2, map,  'size', [3 9])
```

The **reshape** command is used to, well, reshape data arrays and here we used it to simply add one more (empty) dimension so that `montage` will read the data. The result is shown in Fig. 5.8.

We can again inspect the dimensions and data-types using the **whos** command, this time with an argument that restricts the result to any variable beginning with `mriStack`

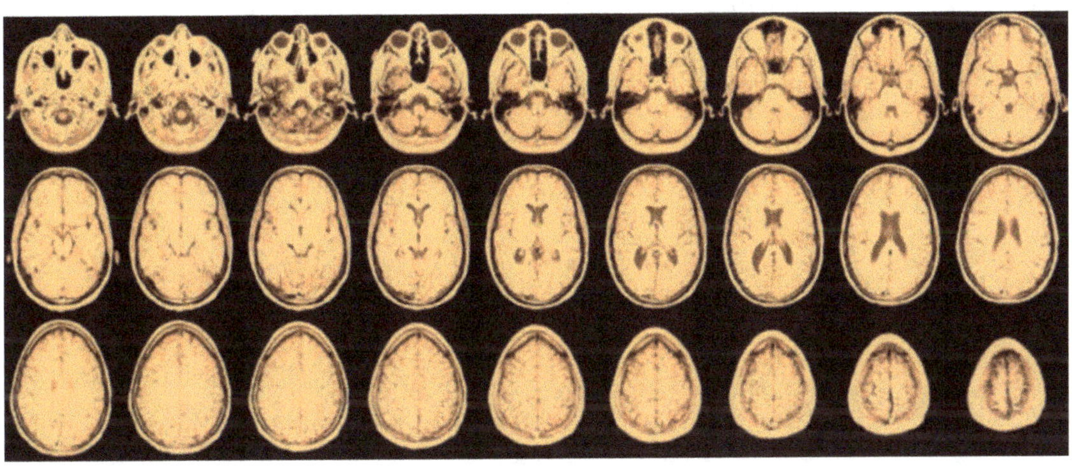

Fig. 5.8 A montage of the 27 images in the MRI stack, arranged as 3×9 and displayed with the colormap "copper"

```
>> whos mriStack*
  Name              Size                 Bytes  Class     Attributes

  mriStack       226x186x27            1134972  uint8
  mriStack2         4-D                1134972  uint8
```

To get the dimensions of the 4D `mristack2` variable we use the command **`size`**

```
> size(mriStack2)

ans =

     226   186     1    27
```

Here, the third dimension is the color channel.

5.4.4 Smoothing, Thresholding and All That

Yes, of course we can perform all these operations and here is a small taste of how it is done. We are going to

1. Load an image and invert it
2. Create a copy of it that has been smoothed with a Gaussian kernel
3. Determine the Otsu threshold for this copy
4. Create a binary image based on the smoothed copy
5. Display the mask on the original
6. Apply this mask (binary image) to the original and make measurements through it
7. Display measurements directly on the original (inverted) image

In the editor, we first initialize, then load, invert, and display the result:

```
%% --- INITIALIZE ---
clear variables
close all
clc
tfs = 16;  %title font size

%% --- load image ---
cd ~/Desktop
blobs       = imread('blobs.tif');  % read tif
blobs_inv   = 255 - blobs;  %invert 8bit image

%% --- display the inverted image ---
figure(1)
imshow(blobs_inv,  'initialmagnification',  'fit')
title('Inverted',  'fontsize', tfs)
```

Next, we smooth the inverted image with a Gaussian kernel, detect the Otsu threshold, apply it, and display the result:

```
%% --- Gaussian smooth and Otsu threshold ---
blobs_inv_gauss = imgaussfilt(blobs_inv, 2);  % sigma = 2 pixels
OtsuLevel       = graythresh(blobs_inv_gauss);  % find threshold
blobs_bw        = imbinarize(blobs_inv_gauss, OtsuLevel); % apply threshold

%% --- display the thresholded image ---
figure(2)
imshow(blobs_bw,  'initialmagnification',  'fit')
title('Inverted, Smoothed, Thresholded',  'fontsize', tfs)
```


To illustrate, on the grayscale image, what we have determined as foreground, we mask it with the binary image `blobs_bw` by multiplying pixel-by-pixel:

```
%% --- mask the inverted image with the thresholded image ---
blobs_bw_uint8   = uint8(blobs_bw);  % convert logical to integer
blobs_masked     = blobs_inv .* blobs_bw_uint8;  % mask image

%% --- display the masked image ---
figure(3)
imshow(blobs_masked,  'initialmagnification',  'fit')
title('Inverted and Masked',  'fontsize', tfs)
```

As an alternative to showing the masked image we can choose to show the outlines of the connected components (the detected blobs):

```
%% --- find perimeter of connected components ---
blobs_perimeter  = bwperim(blobs_bw);   % perimeter of white connected components
blobs_summed   = blobs_inv + 255 * uint8(blobs_perimeter); % convert, scale, and overlay perimeter on image

%% --- display image with perimeter overlaid ---
figure(4)
imshow(blobs_summed,  'initialmagnification',  'fit')
title('Inverted, Masked, Outlines',  'fontsize', tfs)
```

In step two we convert the `logical` variable `blobs_perimeter` to an 8-bit unsigned integer on the fly (and multiplied it by 255 to increase the intensity), before adding it to the image. If you wonder why we do this conversion, just try to omit it and read the error-message from MATLAB.

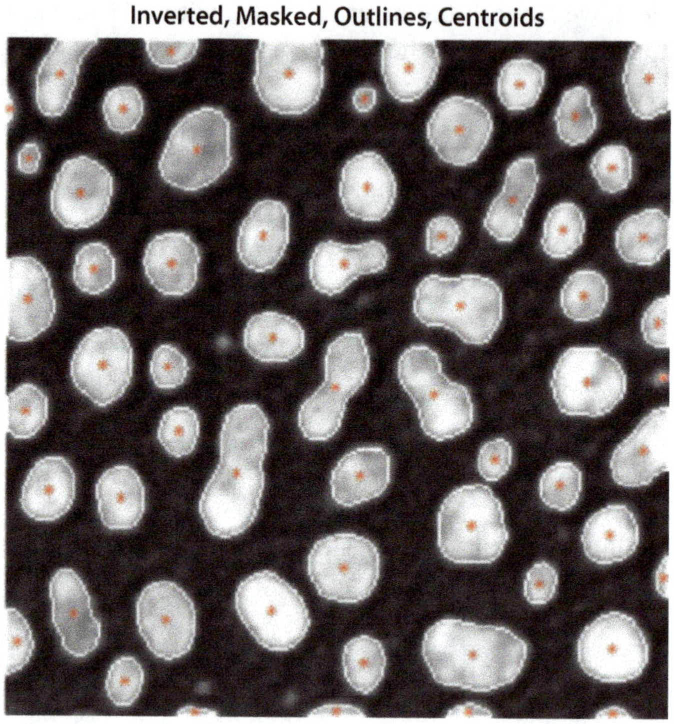

Fig. 5.9 "Blobs" shown with outlines of threshold-based segmentation overlaid. The centroid of each connected component is marked with a red asterisk

Now, let's make some measurements on the b/w image and display them on the `blobs_summed` image from above:

```
%% --- measure areas etc on b/w image ---
stats = regionprops(blobs_bw, 'area', 'perimeter', 'centroid'); % extract features from thresholded image
centroids = cat(1, stats.Centroid); % reformat the centroid data to array

%% --- display centroid positions overlaid on grayscale with out-lines ---
figure(4)  % this figure already exists, we are now adding to it
hold on  % tell MATLAB too keep what is already in the figure
plot(centroids(:, 1), centroids(:, 2), '*r') % use red asteriks
title('Inverted, Masked, Outlines, Centroids', 'fontsize', tfs)
```

The result of this step is shown in Fig. 5.9.

Finally, we measure the gray-scale image using the masks—this should remind you of the "Redirect to:" option in ImageJ (Analyze > Set Measurements …):

```
%% --- measure grayscale values ---
labels          = bwlabel(blobs_bw);  % get identifier for each blob
statsGrayscale   = regionprops(labels, blobs_inv,    'meanInten-sity'); % measure pixel-mean for each blob
```

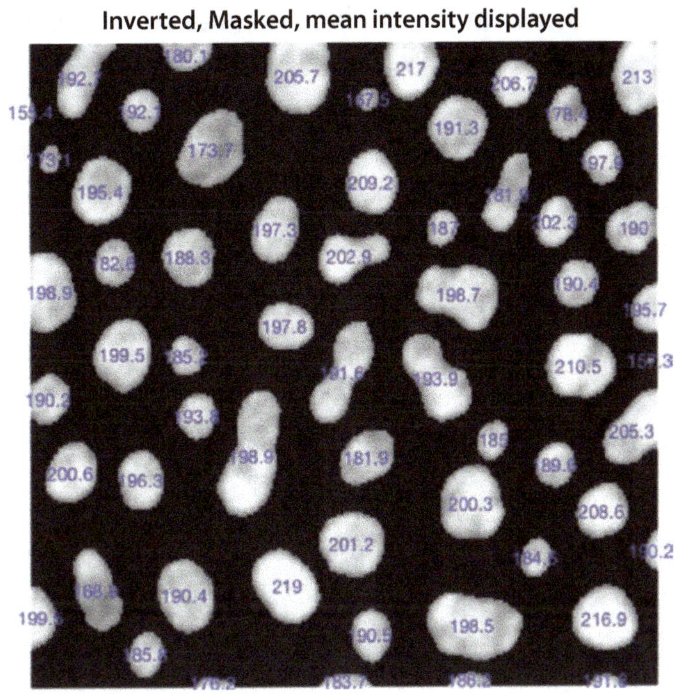

Fig. 5.10 Masked version of "blobs" with the measured mean intensity for each connected component shown


```
meanIntensity   = cat(1, statsGrayscale.MeanIntensity); % reformat the extracted date

%% --- display measurements on image ---
% again, we add to an already existing image
figure(3); hold on
xOffset = 10;  % number of pixels to shift the text to the left
text(centroids(:,   1)   -   xOffset,   centroids(:,   2), num2str(meanIntensity, 4),  'color',  'blue',  'fontsize', 10)
```

Here, we subtracted 10 from the *x*-coordinate to shift the text ten pixels to the left and thereby centering it a bit better on the detected blobs. We also indicate that we want at most four digits displayed.

The result is shown in Fig. 5.10.

Exercise: Do this and understand each step! The code shown above is available in `blobAnalysis.m`.

5.5 Time-Series Analysis

MATLAB has a dedicated data type called simply `timeseries`. We shall not be using this class here as it is too specialized for what we want to do. At a later stage in your research you might find it useful, but be warned that is was developed probably more with the financial sector in mind and may not support quite the kind of analysis you need to perform.

Whether or not you actually have a time-series or simply an ordered list of data often does not matter. Many of the tools are the same but were indeed developed by people doing signal-processing for, e.g., telephone companies, i.e., they worked on actual time-series data.

5.5.1 Simulating a Time-Series of Brownian Motion (Random Walk)

Physical example: Diffusing molecule or bead. A particle undergoing Brownian motion (read Brown's paper Brown et al. (1828), it is delightful!) is essentially performing a random walk: In one dimension, each step is equally likely to be to the right or left. If, in addition, we make the size of the step follow a Gaussian distribution, we essentially have Brownian motion in one dimension, also known as diffusion. Here, we will simplify a bit and set a number of physically relevant constants to one, just to keep the code simpler.

The code for generating the random numbers goes something like this (see entire script of name `simulateAndPlotBrownianMotion.m`):

```
%% --- INITIALIZE ---
clear variables
close all
clc

% --- simulation settings ---
dt          = 1;   % time between recordings
t           = (0   : 1000) * dt;   % time

%% --- GENERATE RANDOM STEPS ---
stepNumber  = numel(t);   % number of steps to take
seed        = 42;   % "seed" for the random number generator
rng(seed);  % reset generator to postion "seed"
xSteps       = randn(1, stepNumber) * sqrt(dt);   % Gaussian distributed steps of zero mean
```

At this stage you do not have to understand the function of the **sqrt**(dt) command—with dt = 1 it is one anyway—it is here because this is how Brownian motion actually scales with time. The `seed` variable and the `rng` command together control the state in which the (pseudo-)random number generator is started—with a fixed value for `seed` we will always produce the same random numbers (take a moment to ponder the meaning of randomness when combined with a computer).

After this, we calculate the positions of the particle and the experimentally determined speeds (we will return to these in detail below):

```
%% --- CALCULATE POSITIONS AND SPEEDS ---
xPos         = cumsum(xSteps);   % positions of particle
varSteps     = var(xSteps);     % variance of step-distribution

xVelocity    = diff(xPos) / dt;   %  "velocities"
xSpeed       = abs(xVelocity);   % "speeds"

meanSpeed    = mean(xSpeed);
stdSpeed     = std(xSpeed);

```

```
%% --- DISPLAY NUMBERS ---
disp(['VAR steps = ' num2str(varSteps)])
disp(['Average speed = ' num2str(meanSpeed)])
disp(['STD speed = ' num2str(stdSpeed)])
```

In the last three lines we used the command **disp** that displays its argument in the command window. It takes as argument variables of many different formats, incl. numerical and strings. Here, we gave it a string variable that was concatenated from two parts, using the [and] operators (other options are to use the commands `cat`, `strcat`, or `horzcat`). The first part is an ordinary string of text in single quotes, the second part is also a string but created from a numeric variable using the command **num2str**.

The other MATLAB commands **cumsum**, **diff**, **mean**, and **std** do what they say and calculate the cumulative sum, the difference, the mean, and the standard deviation, respectively. Look up their documentation, using the `doc` command, for details and additional input arguments.

5.5.2 Plotting a Time-Series

Ok, now let us plot some of these results:

```
%% --- PLOT STEPS VERSUS TIME ---
figure; hold on; clf
plot(t, xSteps, '-', 'color', [0.2 0.4 0.8])
xlabel('Time [AU]')
ylabel('Step [AU]')
title('Steps versus time')
```

The output of these lines, and a similar pair for the positions, is shown in ◘ Fig. 5.11. See the script `simulateAndPlotBrownianMotion.m` to learn how to tweak plot parameters.

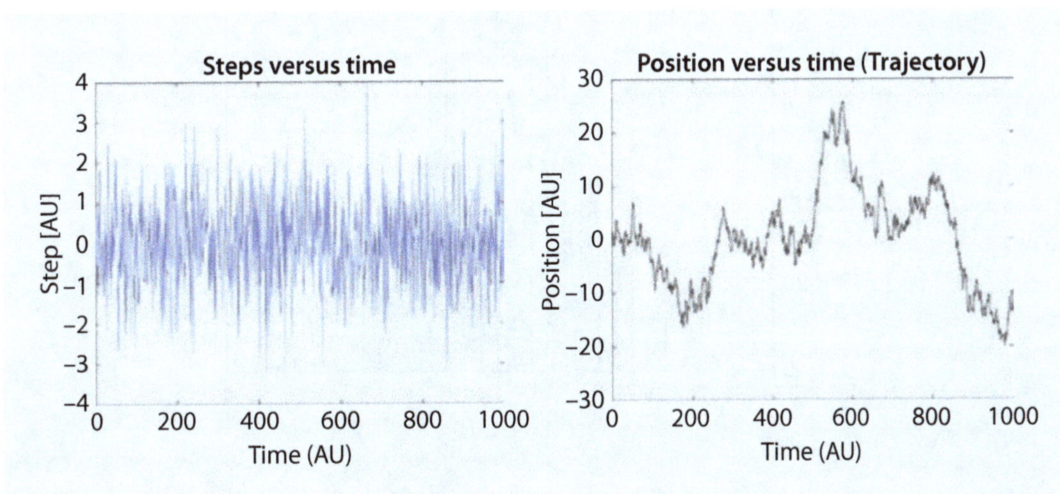

◘ **Fig. 5.11** Steps (left) and positions (right) as a function of time for a one-dimensional random walk

5.5.3 Histograms

Let us now examine the distribution of step-sizes. We do that by plotting a histogram:

```
%% --- PLOT HISTOGRAM OF STEPS ---
figure; hold on
binNumber = floor(sqrt(stepNumber));
histHandle = histogram(xSteps, binNumber)
xlabel('Steps [AU]')
ylabel('Count')
title('Histogram of step-sizes')
```

Figure 5.12 show the resulting plot. The command `histogram` was introduced in MATLAB R2014b and replaces the previous command **hist**—they are largely similar but the new command makes it easier to create pretty figures and uses the color-scheme introduced in MATLAB R2014b: Since version R2014b, MATLAB's new default colormap is called "parula" and replaces the previous default of "jet".

5.5.4 Sub-Sampling a Time-Series (Slicing and Accessing Data)

Sometimes we can get useful information about our time-series by sub-sampling it. An example could be a signal x, that is corrupted by nearest-neighbor correlations: To remove this, simply remove every second data-point, like this:

```
x = 0 : 0.1 : 30;
xSubsampled = x(1 : 2 : end);
```

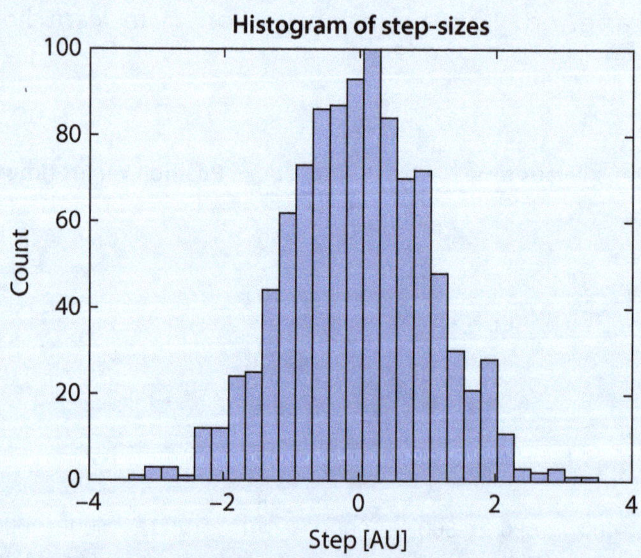

Fig. 5.12 Histogram of step sizes for a random walk. The steps were generated with the command **randn** that creates pseudo-random numbers from a Gaussian distribution

Or, if you wanted only every third data-point from the first 200 entries:

```
xSubsampled = x(1 : 3 : 200);
```

What we just illustrated, was how to read only selected entries from a vector; in the first example we read every second entry from the beginning (the first element in a vector in MATLAB has index 1, not 0), in steps of 2, until the end. The same idea holds for arrays of arbitrary dimension in MATLAB; each dimension is treated independently.

If we wanted, we could also have given a list of indices to read, like this:

```
readThese   = [2 5 7 88 212];  % data-points to read
xSubsampled = x(readThese);
```

Alternatively, if we only wanted to replace a single element, say in entry 7, with the number 3; or find all entries larger than 0.94, then set them to 1:

```
 % --- replace single element ---
x(7)   = 3;  % overwrite/add the 7th element with  "3"

 % --- replace several elements ---
xIndex      = find(x > 0.94);
x(xIndex) = 1;  % write "1" in positions from xIndex
```

The **find** command is very useful for data-wrangling and thresholding. Combined with the query command `isNaN` (asking if something "is not-a-number") you will certainly find yourself applying it once working with real-world data.

5.5.5 Investigating How "Speed" Depends on Δt

After having carefully examined the steps and trajectories we may get the idea of also looking into the velocities and their sizes (speeds). Velocities can be calculated from positions by differentiation wrt. time. Since we have a discrete time-series, we do that by forming the difference and dividing by the time-interval Δt—this is what we did above with the help of the **diff** command.

And this is where it gets interesting: When we vary Δt, our estimate of the speed also changes! Does this make sense? Take a minute to think about it: What we are finding is that, depending on how often we determine the position of a diffusive particle, the estimated speed varies. Would you expect the same behavior for a car or a plane? Ok, if this has you a little confused you actually used to be in good company, that is, until Einstein explained what is really going on, back in 1905—you might know the story.

The take-home message is that speed is ill-defined as a measure for Brownian motion. This is because Brownian motion is a fractal, so, just like when you try to measure the length of Norway's coast-line, the answer you get depends on how you measure. If you are wondering what we can use instead, read on, the next section, on the mean-squared-displacement, has you covered.

5.5.6 Investigating How "Speed" Depends on Subsampling

Another way of investigating the fractal nature of Brownian motion is to directly subsample the already recorded (simulated) time-series of positions. That is, we create a new time-series from the existing one by only reading every second, or third, or fourth etc. time and position data, and then calculate the speed for this new time-series:

```
%% --- SUBSAMPLE THE POSITIONS ---
% --- Re-sample at every fourth time-point ---
t_subsampled_4     = t(1 : 4 : end);
xPos_subsampled_4 = xPos(1 : 4 : end);
meanSpeed4 = mean(abs(diff(xPos_subsampled_4)/dt/4));

% --- Re-sample at every eighth time-point ---
t_subsampled_8     = t(1 : 8 : end);
xPos_subsampled_8 = xPos(1 : 8 : end);
meanSpeed8 = mean(abs(diff(xPos_subsampled_8)/dt/8));
```

Notice how we used, hard to read, compact notation by chaining several commands to calculate the mean speed in a single line—this is possible to do, but usually makes the code harder to read.

Let us now plot these new time-series on top of the original

```
%% --- ZOOMED PLOT SUBSAMPLED POSITION VERSUS TIME ---
figure; hold on;
plot(t, xPos,  '-k.',  'markersize', 16)
plot(t_subsampled_4, xPos_subsampled_4,  '--or',  'markersize', 6)
plot(t_subsampled_8, xPos_subsampled_8,  ':sb',  'markersize', 10)
legend('Every position',  'Every fourth position',  'Every eighth position')
set(gca,  'xlim', [128 152])

xlabel('Time [AU]')
ylabel('Position [AU]')
title('Position versus time')
```

This code, where we added a few extras such as control of the size of markers, should generate a plot like the one shown in Fig. 5.13

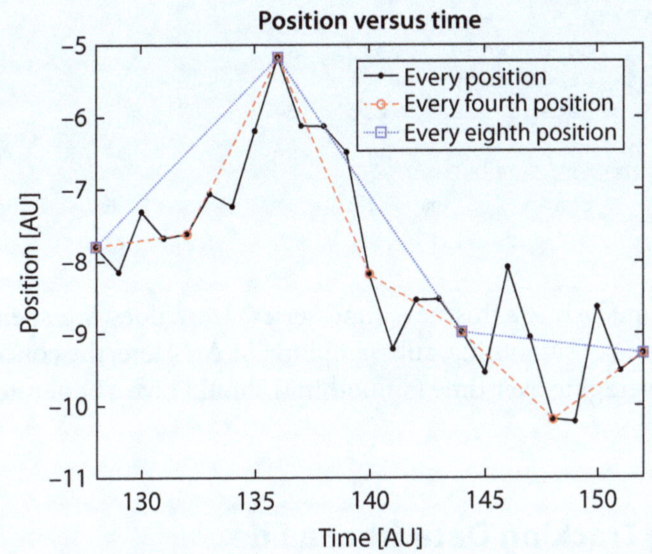

Fig. 5.13 Position as a function of time for a one-dimensional random walk. Black dots show the original time-series. If we had only recorded this trajectory at 1/4 or 1/8 the sampling frequency we would have found the positions indicated by red circles and blue squares, respectively. If we were to estimate the speed for each of these three time-series we would find that the red trace has half ($1/\sqrt{4}$) the speed of the black, and the blue has $1/\sqrt{8} \simeq 0.35$ that of the black. Conclusion: The "speed" depends on how often we measure and is therefore clearly an ill-defined parameter for Brownian motion

5.5.7 Simulating Confined Brownian Motion

Brownian motion doesn't have to be free. The observed particle could be trapped in a small volume or elastically tethered to a fixed point. To be specific, let us choose as physical example a sub-micron sized bead in an optical trap, in water. This turns out to be just as easy to simulate as pure Brownian motion. Writing down the equations of motion and solving them (or using intuition) we see that the observed positions are simply given by random numbers from a Gaussian distribution. The width of the distribution is determined by the strength of the trap (size of the confinement, stiffness of tether). Importantly, we are not sampling the position of this bead very often, only every millisecond or so, rarely enough that it has time to "relax" in the trap between each determination.

```
sampleNumber = 1001;  % number of position determinations
xTrapped     = randn(1, sampleNumber); % position of bead in trap
```

What do we get if we repeat the above analysis? Try it.

5.5.8 Simulating Directed Motion with Random Tracking Error

We may also want to create a time-series that is a hybrid: We have a particle that moves with constant speed in one direction, but the position determination is contaminated with random tracking errors. The simulation, again, is simple:

```
%% --- INITIALIZE ---
dt = 1;  % time between recordings
t  = 0 : dt : 1000 * dt;  % time
v  = 7;  % constant translation speed

%% --- GENERATE POSITIONS ---
xPos = v*t + randn(1, sampleNumber);  % position of bead in trap
```

Repeat the above analysis for this new time-series. How does the speed determination depend on the degree of smoothing, sub-sampling, or Δt? Here, the concept of speed does make sense, and averaging over time (smoothing) should give a better determination, see ▶ Sect. 5.5.10.

5.5.9 Loading Tracking Data from a File

Instead of analyzing simulated data we often want to work on actual experimental data. If your data was generated in ImageJ with the `TrackMate` plugin, the output (when exporting tracks) would be an XML file and we would need a parser (reader) for it called `importtrackmatetracks.m` in order to get the data into MATLAB. See introduction here and code here. This function will return a cell-array of tracks consisting of time, x, y, and z positions (the concept of a **`function`** is explained below in ▶ Sect. 5.6.1.1):

```
function [tracks, metadata] = importTrackMateTracks(file, clipz, scalet)
%%IMPORTTRACKMATETRACKS Import linear tracks from TrackMate
%
% This function reads a XML file that contains linear tracks generated by
% TrackMate (http://fiji.sc/TrackMate). Careful: it does not open the XML
% TrackMate session file, but the track file exported in TrackMate using
% the action 'Export tracks to XML file'. This file format contains less
% information than the whole session file, but is enough for linear tracks
% (tracks that do not branch nor fuse).
%
% SYNTAX
%
% tracks = IMPORTTRACKMATETRACKS(file) opens the track file 'file' and
% returns the tracks in the variable 'tracks'. 'tracks' is a cell array,
% one cell per track. Each cell is made of 4xN double array, where N is the
% number of spots in the track. The double array is organized as follow:
```

```
% [Ti, Xi, Yi, Zi ; ...] where T is the index of the frame the spot has been
% detected in. T is always an integer. X, Y, Z are the spot spatial
% coordinates in physical units.
.
.
.
```


To get a feeling for the data: Pick a few individual tracks and submit them to the same analysis as above. Try a few from the different experimental conditions (try both long and short tracks). Do you notice any difference?

5.5.10 Smoothing (Filtering) a Time-Series

If you suspect that some of the jitter in your signal is simply noise, you can smooth the signal. This is very much the same procedure as when smoothing an image. The relevant command is `smooth` (requires the Curve Fitting Toolbox) and it has several options for adaptation to your needs:

```
% --- simple smoothing ---
xPosSmoothed = smooth(xPos);  % defaults to moving average over 5 data points

% --- sophisticated smoothing ---
span   = 7;             % number of data points to average over
method =  'sgolay' ;    % Savitsky-Golay filter
degree = 3;             % the order of the s-g filter

xPosSmoothed = smooth(xPos, span, method, degree);
```

5.6 MSD: Mean Square Displacement

Motivated by the shortcomings of the speed as a measure for motion, we try our hands at another measure. This measure, while a bit more involved, does not suffer the same problems as the speed but takes a little getting used to. Without further ado:

The mean square displacement for a one-dimensional time-series $x(t)$, sampled continuously, is defined as

$$\text{msd}(\tau) \equiv \langle [x(t+\tau) - x(t)]^2 \rangle, \tag{5.1}$$

where $\langle \cdot \rangle$ is the expectation value of the content, either in the ensemble sense or with respect to t (same thing if the system is ergodic)—think of it as the average over all time-points. It measures how far a particle has moved, in an average sense, in a time-interval of size τ.

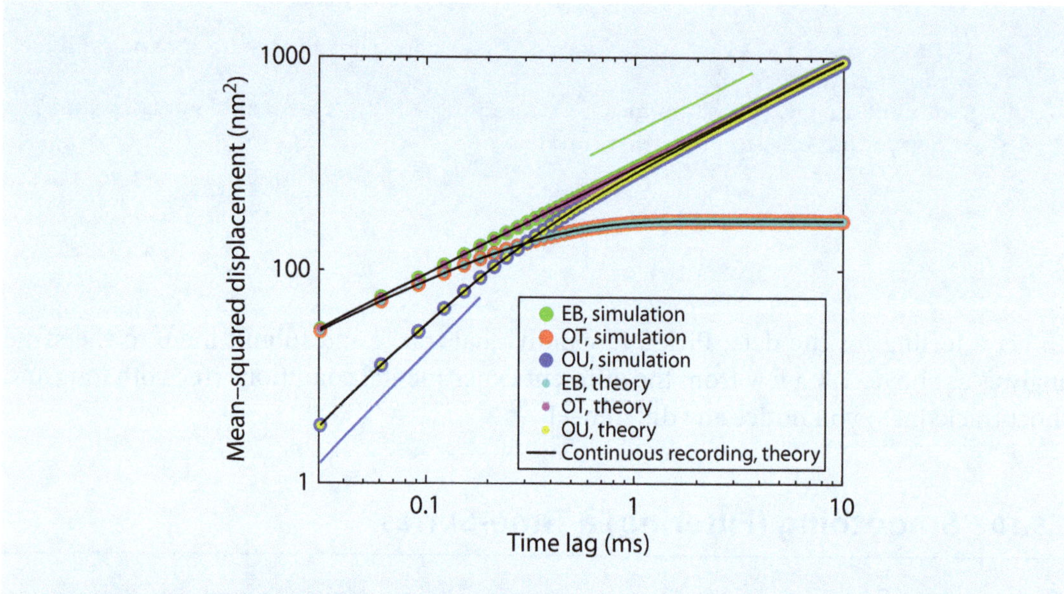

Fig. 5.14 Mean-squared displacement for the Ornstein-Uhlenbeck process (persistent random motion), Brownian motion in an optical trap (confined diffusion), and Brownian-motion proper (free diffusion). Straight lines show slopes of one (green) and two (blue), for comparison to the cases of Brownian motion and linear motion. Green points: Freely diffusing massless particle (Einstein's Brownian motion); red points: trapped massless particle (OT limit, or OU velocity process); and blue points: freely diffusing massive particles (time integral of OU process). This is Fig. 8 in Nørrelykke and Flyvbjerg (2011)

In practice, we need to replace the expectation-value-operation $\langle\cdot\rangle$, with something we can calculate based on our data. There are several ways of doing this, see Qian et al. (1991), and the following is one of the more popular and meaningful ones, for time-lag $\tau = k\,\Delta t$:

$$\text{msd}_k = \frac{1}{M-k}\sum_{M-k}^{n=1}\left[x_{n+k} - x_n\right]^2, \quad k = 1, 2, \ldots M-1 \tag{5.2}$$

where M is the number of postion-determinations of x. Please note, that we are averaging over the track itself using a sliding window: This means that our estimates for the MSD are not independent for consecutive values of the time-lag τ—this is the price we pay for reducing noise and using all the data.

Figure 5.14 shows theoretical and simulated results for the MSD for three different types of motion: (1) Brownian motion (free diffusion); (2) Brownian motion in an optical trap (confined diffusion); and (3) Random motion with finite persistence (Ornstein-Uhlenbeck process)

5.6.1 Creating a Function That Calculates MSDs

One of the great thing about the MSD is that there are no approximations when moving from continuous to discrete time: There are no sampling artifacts. For a fixed time-lag, the MSD can be calculate in MATLAB by defining a function like this:

```
function msd_tau = fun_msd_at_tau_1dim(x,tau)

 % fun_msd_at_tau_1dim FUNCTION
 % GIVEN INPUT DATA 'X' THIS FUNCTION RETURNS THE
 % MEAN-SQUARED-DISPLACEMENT CALCULALTED IN OVERLAPPING WINDOWS
 % FOR THE FIXED TIMELAG VALUE 'tau'
 % NB: THIS IS FOR A SINGLE TIMELAG ONLY BUT AVERAGED OVER THE ENTIRE TRACK

 % 2016-06-03, sfn, created
 % 2016-06-10, sfn, modified for one dimension
 % 2017-05-15, sfn, nomenclature changes

 %% --- INITIALIZE ---
 M   = length(x);              % number of postions determined
 dr2  = zeros(1, M - tau);     % initialize and speed up procedure

  %% --- CALCULATE THE MSD AT A SINGLE TIMELAG ---
 for k = 1 : M - tau
     dx2      = (x(k + tau) - x(k)).^2;  % squared x-displacement

     dr2(k) = dx2;  % store the squared x-displacement for each position of the sliding window
 end

 msd_tau       = mean(dr2);   % The mean of the squared displacements calculated in sliding windows
```

In this code-example you should notice that we declared a **function**, used the command **zeros** to pre-allocate memory hence speed up the procedure, and squared each element in a vector with the .^ operator which should not be confused with the ^ operator that would have attempted to form the inner product of the vector with itself (and fail). If your function-call fails, you might have to tell MATLAB where to find the function using the `addpath` command or by clicking on "set path" in the HOME tab and then pointing to the folder that holds the function.

5.6.1.1 About Functions and How to Call Them

A function is much like a normal script except that it is blind and mute: I doesn't see the variables in your workspace and whatever variables are defined inside of the function are not visible from the workspace either. One way to get data into the function is to feed it explicitly as input, here as `x` and `tau`. The only data that gets out is that explicitly stated as output, here `msd_tau`. This is how you call the function `msd_tau`, ask it to calculate the mean square displacement for the time-series with coordinates (x, y), for a single time-lag of $\tau = 13$ and return the result in the variable `dummy`:

```
>> dummy = fun_msd_at_tau_1dim(x, 13);
```

Having learnt how to do this for a single time-lag, we can now calculate the MSD for a range of time-lags using a **for** loop:

```
for tau = 1 : 10
  msd(tau) = fun_msd_at_tau_1dim(x, tau);
end
```

After which we will have a vector of length ten holding the MSD for time-lags one through ten. If the physical time-units are non-integers you simply plot MSD against these, do not try to address non-integer positions in a vector or matrix, they do not exist. This will become clear the first time you try it.

To build some further intuition for how the MSD behaves, let us calculate it analytically for a couple of typical motion patterns.

5.6.2 MSD: Linear Motion

By linear motion we mean

$$x(t) = vt, \tag{5.3}$$

where v is a constant velocity and t is time. That is, the particle was at position zero at time zero, $x(t=0)=0$, and moves to the right with constant speed. The MSD then becomes

$$\mathrm{msd}(\tau) = \langle [vt + v\tau - vt]^2 \rangle = v^2\tau^2, \tag{5.4}$$

i.e., the MSD grows with the square of the time-lag τ. In a double-logarithmic (log-log) plot, the MSD would show as a straight line of slope 2 when plotted against the time-lag τ:

$$\log \mathrm{msd}(\tau) = \log v^2 + 2\log\tau \tag{5.5}$$

5.6.3 MSD: Brownian Motion

By Brownian motion we mean

$$\dot{x}(t) = a\eta(t), \tag{5.6}$$

where $\cdot$ means differentiation wrt. time, $a = \sqrt{2D}$, D is the diffusion coefficient and η is a normalised, Gaussian distributed, white noise

$$\langle \eta(t) \rangle = 0, \quad \langle \eta(t)\eta(t') \rangle = \delta(t - t'), \tag{5.7}$$

where δ is Dirac's delta function. See Wikipedia for an animation of Brownian motion:

▶ https://en.wikipedia.org/wiki/Brownian_motion

With this equation of motion we can again directly calculate the MSD:

$$\mathrm{msd}(\tau) = \langle\left[\int_{-\infty}^{t+\tau} \mathrm{d}t' \dot{x}(t') - \int_{-\infty}^{t} \mathrm{d}t' \dot{x}(t')\right]^2\rangle \tag{5.8}$$

$$= a^2 \tau = 2D\tau, \tag{5.9}$$

a result that should be familiar to some of you.

Apart from prefactors, that we do not care about here, the crucial difference is that the MSD now grows linearly with the time-lag τ, and in a log-log plot it would hence be a straight line with slope one when plotted against τ.

5

We are much more interested in the mathematical properties of this motion than in the actual thermal self-diffusion coefficient D: The temporal dynamics of this equation can be used to model systems that move randomly, even if not driven by thermal agitation. So, when we say Brownian motion, from now on, we mean the mathematical definition, not the physical phenomenon.

For those interested in some mathematical details, Brownian motion can be described via the Wiener process W, with the white noise being the time-derivative of the Wiener process $\eta = \dot{W}$. The Wiener process is a continuous-time stochastic process and is one of the best known examples of the broader class of Levý processes that can have some very interesting characteristics such as infinite variance and power-law distributed step-sizes. These processes come up naturally in the study of the field of distributions, something you can think of as being a generalization of ordinary mathematical functions, and also requires an extension of normal calculus to what is known as Itô calculus. If you are into mathematical finance or stochastic differential equations you will know all of this already.

5.6.3.1 MSD: Simulated Random Walk

We can also calculate the MSD for the discrete random walk that we simulated earlier. There, we simplified our notation by setting $2D=1$ but otherwise the random walk was a mathematically exact representation of one-dimensional free diffusion. Here is the calculation, for a time-lag of $\tau = k\Delta t$ and explicitly including the $\sqrt{2D}$ prefactor; you should already have all the ingredients to understand each step:

$$\mathrm{msd}_k = \langle\left[x_{n+k} - x_n\right]^2\rangle \tag{5.10}$$

$$= \langle\left[\sum_{n+k}^{i=1} \Delta x_i - \sum_{n}^{i=1} \Delta x_i\right]^2\rangle = \langle\left[\sum_{n+k}^{i=1+n} \Delta x_i\right]^2\rangle \tag{5.11}$$

$$= \langle\left[\sum_{n+k}^{i=1+n} \zeta_i \sqrt{2D\,\Delta t}\right]^2\rangle = 2D\,\Delta t\langle\left[\sum_{n+k}^{i=1+n} \zeta_i \sum_{n+k}^{j=1+n} \zeta_j\right]\rangle \tag{5.12}$$

$$= 2D\Delta t \sum_{n+k}^{i=1+n} \langle\zeta_i^2\rangle = 2D\,k\Delta t = 2D\,\tau \tag{5.13}$$

here we used that the position at time $n\,\Delta t$ is the sum of the steps before then:

$$x_n = \sum_{n}^{i=1} \Delta x_i, \quad \Delta x_i = \zeta_i \sqrt{2D\,\Delta t} \tag{5.14}$$

where ζ are Gaussian distributed random numbers of zero mean, unit variance, and uncorrelated:

$$\langle \zeta_i \zeta_j \rangle = \delta_{i,j}, \quad \langle \zeta_i \rangle = 0 \tag{5.15}$$

with $\delta_{i,j}$ Kronecker's delta: Zero for i and j different, unity if they are the same. These ζ-values are the ones we created with the **randn** command in MATLAB. Again, we see that the MSD is linear in the time-lag $\tau = k\,\Delta t$.

5.6.4 MSD: Averaged Over Several 2-Dim Tracks

To start quantifying the motion of multiple tracks, in two spatial dimensions, we first calculate the mean-squared-displacement for an individual track m

$$\mathrm{msd}_{k,m} = \frac{1}{M_m - k} \sum_{M_m - k}^{i=1} \left((x_{i+k} - x_i)^2 + (y_{i+k} - y_i)^2 \right), \tag{5.16}$$

where $k = 1, 2, \ldots, M_m - 1$ is the time-lag in units of Δt and M_m is the number of positions determined for track m. Notice, that we use a sliding window so that the $M_m - k$ determinations of the MSDs at time-lag $k\,\Delta t$ are not independent; this introduces strong correlations between the MSD calculated at neighboring time-lags by trading independence for smaller error-bars Wang et al. (2007).

One way to calculate the sample-averaged MSD is to weigh each MSD by the number of data-points used to calculated it

$$\mathrm{MSD}_k = \frac{1}{\Sigma_m (M_m - k)} \sum_m (M_m - k)\,\mathrm{msd}_{k,m}, \tag{5.17}$$

where the sums extend over all time-series with $M_m > k$. Here, the weights are chosen as equal to the number of intervals that was used to calculate the MSD for a given time-lag and track.

5.6.5 Further Reading About Diffusion, the MSD, and Fitting Power-Laws

Papers dealing with calculation of the MSD: Qian et al. (1991) and under conditions with noise: Michalet (2010). Analytically exact expressions for several generic dynamics cases (free diffusion, confined diffusion, persistent motion both free and confined): Nørrelykke and Flyvbjerg (2011). Determining diffusion coefficients when this or that moves or not, this is an entire PhD thesis compressed to one long paper: Vestergaard et al. (2014). How to fit a power-law correctly and what can happen if you do it wrong like most people do—*an absolute must-read:* Clauset et al. (2009).

Take Home Message

Ok, good, you made it to here. Congratulations!

If this was your first encounter with coding, MATLAB, or numerical simulations you may feel a bit overwhelmed at this point—don't worry, coding isn't mastered in one day; put in the hours and you will learn to master MATLAB, like many have before you. If you already knew MATLAB you probably skipped this chapter.

Here is what you just learned how to do in MATLAB:

- Create a plot and save it to a file in pdf, png, or other formats
- Load an image and process it (smoothing, thresholding)
- Perform measurements on an image and overlay those measurements on the image
- Read and modify the meta-data in an image file
- Simulate a random walk as a model for free diffusion (Brownian motion), confined/tethered motion, and directed motion with tracking error
- Calculate and display the mean square displacement (MSD)—a robust measure of motion
- Spot when "speed" is a flawed measure for motion (the mean will depend on the sampling interval)—when there is a random component, it is always a flawed measure
- Structure and document your code, keeping good code hygiene

Acknowledgements We thank Ulrike Schulze (Wolfson Imaging Centre–MRC Weatherall Institute of Molecular Medicine, Micron Oxford Advanced Bioimaging Unit, University of Oxford) for reviewing this chapter.

Appendix: MATLAB Fundamental Data Classes

All data stored in MATLAB has an associated class. Some of these classes have obvious names and meanings while others are more involved, e.g. the number `12` is an integer, whereas the number `12.345` is not (it is a `double`), and the data-set `{12, 'Einstein', 7+6i, [1 2 ; 3 4]}` is of the class `cell`. A short video (5min) about MATLAB fundamental classes and data types.

Here are some of the classes that we will be using, sometimes without needing to know it, and some that we won't:

single, double - 32 and 64 bit floating number, e.g. `1'234.567` or `-0.000001234`. The default is `double`.

int8/16/32/64, uint8/16/32/64 - (unsigned-)integers of 8/16/32/64 bit size, e.g. `-2` or `127`

logical - Boolean/binary values. Possible values are `TRUE, FALSE` shown as `1, 0`

char - characters and strings (largely the same thing), e.g. `'hello world!'`. Character arrays are possible (all rows must be of equal length) and are different from cell arrays of characters.

cell - cell arrays. For storing heterogeneous data of varying types and sizes. Very flexible. Great potential for confusion. You can have cells nested within cells, nested within cells …

struct - structure arrays. Like cell arrays but with names and more structure; almost like a spreadsheet.

table - tables of heterogeneous but tabular data: Columns must have the same number of rows. Think "spreadsheet". Supports useful commands such as `summary`. *New data format from 2013b.*

categorical - categorical data such as `'Good', 'Bad', 'Horrible'`, i.e., data that take on a discrete set of possible values. Plays well with `table`. *New data format from 2013b.*

MATLAB Documentation Keywords for Data Classes

The following is a list of search terms related to the `cell`, `struct`, and `table` data classes. They are titles of individual help-documents and are provided here because the documentation of MATLAB is vast and it can take some time to find the relevant pages. Simply copy and paste the lines into MATLAB's help browser in the program or on the web

```
Access Data in a Cell Array
Cell Arrays of Character Vectors
Multilevel Indexing to Access Parts of Cells
Access Data in a Structure Array
Cell vs. Struct Arrays
Create and Work with Tables
Access Data in a Table
```

Here is a link to a video about tables and categorical arrays.

Appendix: Do I Have That Toolbox?

To find out which toolbox a particular command requires simply search for it in the documentation and notice the path. Alternatively, use the **`which`** command:

```
>>  which('graythresh')
/Applications/MATLAB_R2015b.app/toolbox/images/images/
graythresh.m
```

Any path to a function, as found with the **`which`** command, that includes `.../toolbox/matlab/...` does not require a specific toolbox as it is part of the core MATLAB distribution. It is also possible to use the `matlab.codetools.requiredFilesAndProducts` command:

```
>> [fileList,productList] = matlab.codetools.requiredFilesAndProd-
   ucts('graythresh');
>> productList.Name

ans =
MATLAB

ans =
Image Processing Toolbox
```

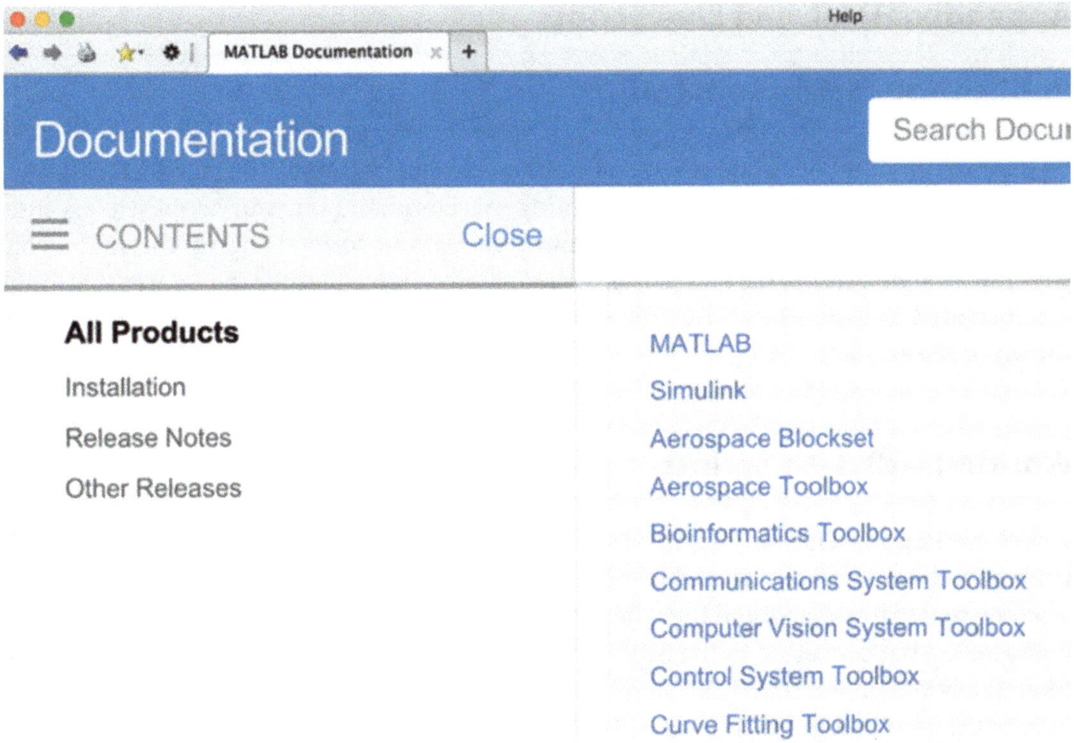

Fig. 5.15 Truncated view of Help window showing "All Products" which includes all the toolboxes you have installed—if it doesn't say "toolbox" it is something else

To find out which toolboxes you have installed, say `doc` to start the help-browser and click "All Products", see Fig. 5.15.

Alternatively, navigate to the folder where MATLAB is installed, via command line or MATLAB or Finder. Example for an installation on a Mac, getting the list in iTerm (bash):

```
[simon@SimonProRetina ~]$ ls -lho /Applications/MATLAB_R2015b.app/
toolbox/ | head
total 0
drwxr-xr-x    5 simon   170B Oct 13  2015 aero
drwxr-xr-x    7 simon   238B Oct 13  2015 aeroblks
drwxr-xr-x   12 simon   408B Oct 13  2015 bioinfo
drwxr-xr-x   23 simon   782B Oct 13  2015 coder
drwxr-xr-x   11 simon   374B Oct 13  2015 comm
drwxr-xr-x   40 simon   1.3K Oct 13  2015 compiler
drwxr-xr-x    8 simon   272B Oct 13  2015 compiler_sdk
drwxr-xr-x    8 simon   272B Oct 13  2015 control
drwxr-xr-x    8 simon   272B Oct 13  2015 curvefit
[simon@SimonProRetina ~]$
```

Here, you need to be able to recognize that the toolbox names are abbreviated, so that, e.g., the `Aerospace Toolbox` is referred to simply as `aero`.

Appendix: HTML and Live Scripts

Publish Your Script to HTML

If you want to show your code and it's output to someone, without running MATLAB, you can do it with the PUBLISH feature. Running this command on your script will execute it and create a folder called "html" in the same place as your script. Inside of this folder you will find a single .html file and perhaps a number of .png files for the figures that your script created. Figure 5.16 shows the result of publishing to HTML the same code as was shown in Fig. 5.4.

Working with Live Scripts

Live Script is a new feature in MATLAB R2016a. You can think of it as something in between publishing to HTML and working directly in the script editor. Existing scripts can be converted to live scripts and the other way around! Figure 5.17 shows the same code as in Fig. 5.4, but converted to the live script format (extension mlx). If you have seen iPython notebooks or Mathematica you might see what the inspiration is. Try it out, you might like it! Just keep in mind that it is a new feature and that you cannot share your live-scripts with anyone using an older version than R2016a (unless you convert to standard m-file first).

Appendix: Getting File and Folder Names Automatically

Read from a Folder

To get a list of files in a folder you have several options: (1) Navigate MATLAB to the folder (by clicking or using the **cd** command) and type ls or dir; (2) Give the **ls** or **dir** command followed by the path to the folder, like this

```
>> ls /Users/simon/Desktop/
blobs.tif  mri-stack.tif

>> dir /Users/simon/Desktop/

.               blobs.tif
..              mri-stack.tif
```

We can also assign the output to variables:

```
>> lsList = ls('/Users/simon/Desktop/');
>> dirList = dir('/Users/simon/Desktop/');
```

What is the difference between the two variables dirList and lsList?

Contents


```
% myFirstScript.m

% This script demonstrates documentation and sections

% 2016-06-02, sfn: created
% 2016-06-03, sfn: added plot lines
% 2016-03-21, sfn: Show value of "A" and "stepsize"
```

--- INITIALIZE ---

```
clear variables
close all
clc
A           = 10   % The peak-amplitude
stepsize    = 0.01 % the granularity of the the x-vector
maxX        = 10;    % the maximum value x can take
x           = 0 : stepsize : maxX; % creating the x-vector
```

```
A =

    10

stepsize =

    0.0100
```

--- FUNCTIONS OF X ---

```
y   = A * cos( x ); % a simple cosine
y2  = y .* x;       % a cosine with linearly growing amplitude
```

--- PLOTS ---

```
figure, hold on, box on
plot( x, y )
plot( x, y2, '--r' )
legend('cos(x)', 'cos(x)/x')
xlabel( 'Time (AU)' )
ylabel( 'Position (AU)' )
title( 'Plots of various sinusoidal functions' )
```

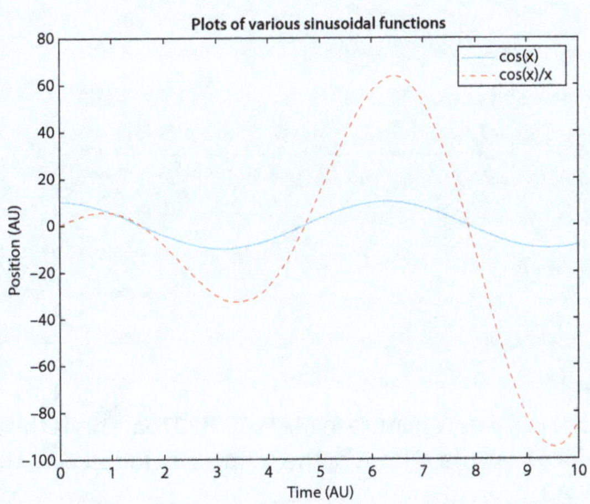

--- save to file ---

```
cd ~/Desktop
% print( '-dpdf', 'cosineFigure.pdf' )
```

Published with MATLAB® R2016a

Fig. 5.16 Example of publishing code to HTML. This is the same code as in the m-script shown in Fig. 5.4. Notice how the output of the script is included with the code. This is an HTML file and therefor easy to share, but you cannot execute it in MATLAB

```
% myFirstScript.m

% This script demonstrates documentation and sections

% 2016-06-02, sfn: created
% 2016-06-03, sfn: added plot lines
% 2016-03-21, sfn: converted to live-script. Show value of "A" and "stepsize"
```

--- INITIALIZE ---

```
clear variables
close all
clc
A           = 10   % The peak-amplitude

 A = 10

stepsize    = 0.01 % the granularity of the the x-vector

 stepsize = 0.0100

maxX        = 10;    % the maximum value x can take
x           = 0 : stepsize : maxX; % creating the x-vector
```

--- FUNCTIONS OF X ---

```
y   = A * cos( x ); % a simple cosine
y2  = y .* x;        % a cosine with linearly growing amplitude
```

--- PLOTS ---

```
figure, hold on, box on
plot( x, y )
plot( x, y2, '--r' )
legend('cos(x)', 'cos(x)/x')
xlabel( 'Time (AU)' )
ylabel( 'Position (AU)' )
title( 'Plots of various sinusoidal functions' )
```

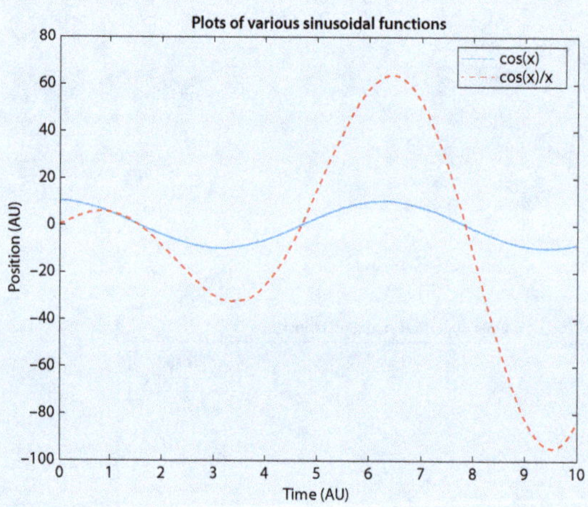

--- save to file ---

```
cd ~/Desktop
% print( '-dpdf', 'cosineFigure.pdf' );
```

Fig. 5.17 Example of `Live Script`, a new feature in MATLAB R2016a. This is the same code as in the `m`-script shown in Fig. 5.4. Notice how the output of the script is included with the code. This script can be edited and executed in MATLAB

Path and File Names

To illustrate how to work with and combine file-names and path-names we will introduce the dialogue window (again assuming we are in the `/Users/simon/Desktop/` directory and have a file called "blobs.tif" there):

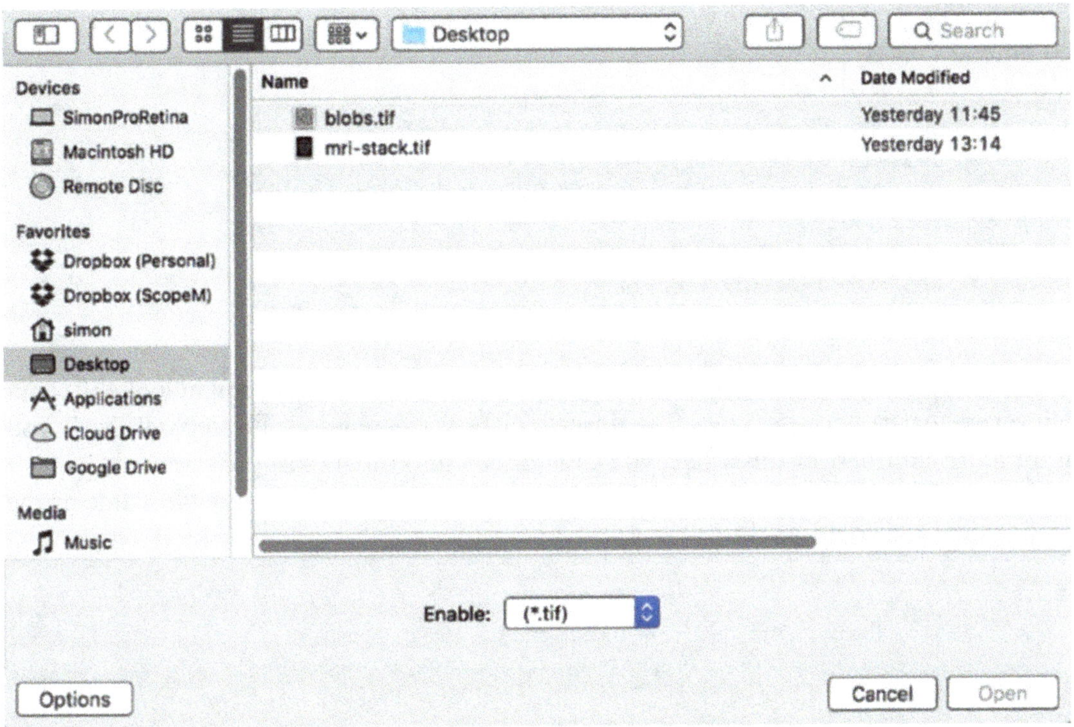

Fig. 5.18 The dialogue window, in OS X, that appears in response to the **uigetfile** command

```
>> fileName = uigetfile('.tif')
```

In response to which we should see a dialogue window similar to Fig. 5.18.

We should also be told the name of the file we selected:

```
fileName =

blobs.tif
```

If we want more information, such as the location of the file we do:

```
>> [fileName, pathName] = uigetfile('.tif')

fileName =

blobs.tif

pathName =

/Users/simon/Desktop/
```

From the file- and path-name we can now create the full file-name, incl. the path, using the command `fullfile`:

```
>> fullFileName = fullfile(pathName, fileName)

fullFileName =

/Users/simon/Desktop/blobs.tif
```

Obviously, if you are working on a different system you file-separator might look different. However, that is because `fullfile` inserts platform-dependent file separators. If you want more control over this aspect you should look into the `filesep` command.

Reversely, if you had the full name of a file and wanted to extract the file-name or the path-name, you could do this:

```
>> [pathstr,name,ext] = fileparts(fullFileName)

pathstr =

/Users/simon/Desktop

name =

blobs

ext =

.tif
```

Alternatively, if all we wanted was the name of a directory we would use the command `uigetdir`—you can guess what it does.

Why did we just do all this? We did it because we often have to spend a lot of time on data-wrangling before we can even get to the actual data-analysis. Knowing how to easily extract file and path names for your data allows you automate many later steps. Example: You might want to open each image in a directory, crop it, scale it, smooth it, then save the results to another directory with each modified image given the same name as the original but with "_modified" appended to the name.

Appendix: Codehygiene

It is important for your future self, not to mention collaborators, that you keep good practices when coding.

- The actual code should be easy to read, not necessarily as compact as possible
- Use descriptive names
- Document the code
- Insert plenty of blank spaces: Let your code breathe!

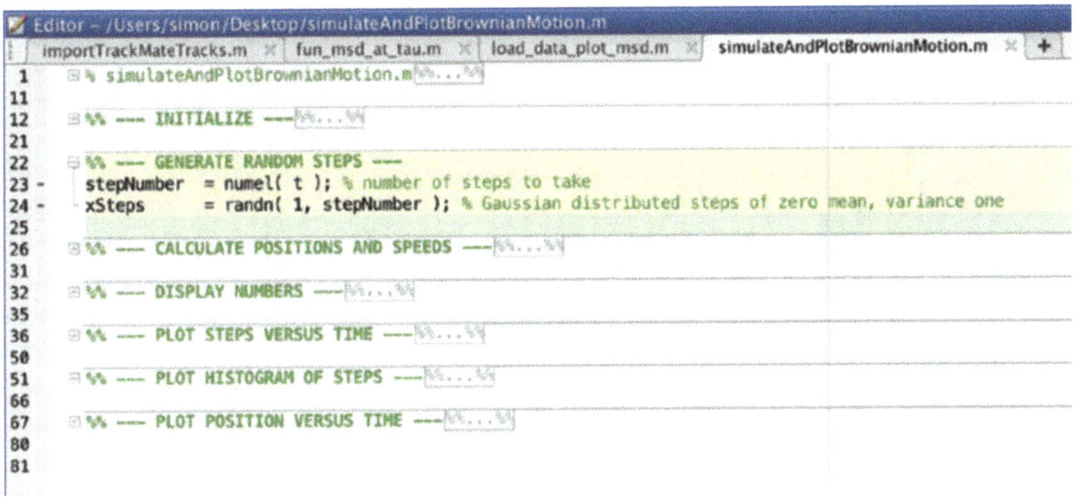

Fig. 5.19 Screenshot of code that is clearly structured and folded. The currently active code-block is highlighted in yellow

Figure 5.19 is and example of how your code could look, when folded, if you take care to structure it nicely—notice how easy it to figure out what goes on where, without having to read a single line of actual code.

Appendix: MATLAB Cheat Sheet

Here are two compact pages that you are encouraged to print separately and keep around when using MATLAB—at least initially. They outline most of the syntax and also list the most commonly used commands. This version (you can find several online) was compiled by Thor Nielsen (thorpn86@gmail.com) ▶ http://www.econ.ku.dk/pajhede/.

Matlab Cheat Sheet

Some nifty commands

clc	Clear command window
clear	Clear system memory
clear x	Clear x from memory
commandwindow	open/select commandwindow
whos	lists data structures
whos x	size, bytes, class and attributes of x
ans	Last result
close all	closes all figures
close(H)	closes figure H
winopen(pwd)	Open current folder
class(obj)	returns objects class
save filename	saves all variables to .mat file
save filename x,y	saves x,y variables to .mat file
save -append filename x	appends x to .mat file
load filename	loads all variables from .mat file
ver	Lists version and toolboxes
beep	Makes the beep sound
doc function	Help/documentation for function
docsearch string	search documentation
web google.com	opens webadress
inputdlg	Input dialog box
methods(A)	list class methods for A

Statistical commands

distrnd	random numbers from dist
distpdf	pdf from dist
distcdf	cdf dist
distrnd	random numbers from dist
hist(x)	histogram of x
histfit(x)	histogram and

*Standard distributions (dist): norm, t, f, gam, chi2, bino
*Standard functions: mean,median,var,cov(x,y),corr(x,y),
*quantile(x,p) is not textbook version.
(It uses interpolation for missing quantiles.

Keyboard shortcuts

edit filename	Opens filename in editor
Alt	Displays hotkeys
F1	Help/documentation for highlighted function
F5	Run code
F9	Run highlighted code
F10	Run code line
F11	Run code line, enter functions
Shift+F5	Leave debugger
F12	Insert break point
Ctrl+Page up/down	Moves between tabs
Ctrl+shift	Moves between components
Ctrl+C	Interrupts code
Ctrl+D	Open highlighted codes file
Ctrl+ R/T	Comment/uncomment line
Ctrl+N	New script
Ctrl+W	Close script
Ctrl+shift+d	Docks window
Ctrl+shift+u	Undocks window
Ctrl+shift+m	max window/restore size

Built in functions/constants

abs(x)	absolute value
pi	3.1415...
inf	∞
eps	floating point accuracy
1e6	10^6
sum(x)	sums elements in x
cumsum(x)	Cummulative sum
prod	Product of array elements
cumprod(x)	cummulative product
diff	Difference of elements
round/ceil/fix/floor	Standard functions..

*Standard functions: sqrt, log, exp, max, min, Bessel
*Factorial(x) is only precise for $x < 21$

Cell commands

A cell can contain any variable type.

x=cell(a,b)	a ×b cell array
x{n,m}	access cell n,m
cell2mat(x)	transforms cell to matrix
cellfun('fname',C)	Applies fname to cells in C

cellfun

Strings and regular expressions

strcomp	compare strings (case sensitive)
strcompi	compare strings (not case sensitive)
strncomp	as strcomp, but only n first letters
strfind	find string within a string , gives start position
regexp	Search for regular expression

Logical operators

- && — Short-Circuit AND.
- & — AND
- || — Short-Circuit or
- | — or
- ~ — not
- == — Equality comparison
- ~= — not equal
- isa(obj, 'class_name') — is object in class

*Other logical operators: <,>,>=,<=
*All above operators are elementwise
*Class indicators: isnan, isequal, ischar, isinf, isvector , isempty, isscalar, iscolumn
*Short circuits only evaluate second criteria if first criteria is passed, it is therefore faster.
And useful fpr avoiding errors occuring in second criteria
*non-SC are bugged and short circuit anyway

Variable generation

j:k	row vector [j,j+1,...,k]
j:i:k	row vector [j,j+i,...,k],
linspace(a,b,n)	n points linearly spaced and including a and b
NaN(a,b)	a×b matrix of NaN values
ones(a,b)	a×b matrix of 1 values
zeros(a,b)	a×b matrix of 0 values
meshgrid(x,y)	2d grid of x and y vectors
[a,b]=deal(NaN(5,5))	declares a and b
global x	gives x global scope

Tables

T=table(var1,var2,...,varN)	Makes table*
T(rows,vars)	get sub-table
T{rows,vars}	get data from table
T.var or T.(varindex)	all rows of var
T.var(rows)	get values of var from rows
summary(T)	summary of table
T.var3(T.var3>5)=5	changes some values
T.Properties.Varnames	Variable names
T = array2table(A)	! make table from array
T = innerjoin(T1,T2)	innerjoin
T = outerjoin(T1,T2)	outerjoin !

Rows and vars indicate rows and variables.
tables are great for large datasets, because they use less memory and allow faster operations.
*rowfun is great for tables, much faster than eg. looping

matrix and vector operations/functions

x=[1, 2, 3]	1x3 (Row) vector
x=[1; 2; 3]	3x1 (Column) vector
x=[1, 2; 3, 4]	2x2 matrix
x(2)=4	change index value nr 2
x(:)	All elements of x (same as x)
x(j:end)	j'th to last element of x
x(2:5)	2nd to 5th element of x
x(j,:)	all j row elements
x(:,j)	all j column elements
diag(x)	diagonal elements of x
x.*y	Element by element multiplication
x./y	Element by element division
x+y	Element by element addition
x-y	Element by element subtraction
A^n	normal/Matrix power of A
A.^n	Elementwise power of A
A'	Transpose
inv(A)	Inverse of matrix
size(x)	Rows and Columns
eye(n)	Identity matrix
sort(A)	sorts vector from smallest to largest
eig(A)	Eigenvalues and eigenvectors
numel(A)	number of array elements
x(x>5)=0	change elemnts >5 to 0
x(x>5)	list elements >5
find(A>5)	Indices of elements >5
find(isnan(A))	Indices of NaN elements
[A,B]	concatenates horizontally
[A;B]	concatenates vertically

For functions on matrices, see bsxfun,arrayfun or repmat
*if arrayfun/bsxfun is passed a gpuArray, it runs on GPU.
*Standard operations: rank,rref,kron,chol
*Inverse of matrix inv(A) should almost never be used, use RREF through \ instead: $inv(A)b = A\backslash b$.

Plotting commands

`fig1 = plot(x,y)`	2d line plot, handle set to fig1
`set(fig1, 'LineWidth', 2)`	change line width
`set(fig1, 'LineStyle', '-')`	dot markers (see *)
`set(fig1, 'Marker', '.')`	marker type (see *)
`set(fig1, 'color', 'red')`	line color (see *)
`set(fig1, 'MarkerSize', 10)`	marker size (see *)
`set(fig1, 'FontSize', 14)`	fonts to size 14
`figure`	new figure window
`figure(j)`	graphics object j
`get(j)`	returns information graphics object j
`gcf(j)`	get current figure handle
`subplot(a,b,c)`	Used for multiple figures in single plot
`xlabel('\mu line','FontSize',14)`	names x/y/z axis
`ylim([a b])`	Sets y/x axis limits for plot to a-b
`title('name','fontsize',22)`	names plot
`grid on/off;`	Adds grid to plot
`legend('x','y','Location','Best')`	adds legends
`hold on`	retains current figure when adding new stuff
`hold off`	restores to default (no hold on)
`set(h,'WindowStyle','Docked');`	Docked window style for plots
`datetick('x',yy)`	time series axis
`plotyy(x1,y1,x2,y2)`	plot on two y axis
`refreshdata`	refresh data in graph if specified source
`drawnow`	do all in event queue

```
* Some markers:  ', +, *, x, o, square
* Some colors:   red, blue, green, yellow, black
* color shortcuts:   r, b, g, y, k
* Some line styles:   -, --, :, -.
* shortcut combination example:   plot(x,y,'b--o')
```

Output commands

`format short`	Displays 4 digits after 0
`format long`	Displays 15 digits after 0
`disp(x)`	Displays the string x
`disp(x)`	Displays the string x
`num2str(x)`	Converts the number in x to string
`num2str(['nA is = ' num2str(a)])`	OFTEN USED! !
`mat2str(x)`	Converts the matrix in x to string
`int2str(x)`	Converts the integer in x to string
`sprintf(x)`	formated data to a string

System commands

`addpath(string)`	adds path to workspace
`genpath(string)`	gets strings for subfolders
`pwd`	Current directory
`mkdir`	Makes new directory
`tempdir`	Temporary directory
`inmem`	Functions in memory
`exit`	Close matlab
`dir`	list folder content
`ver`	lists toolboxes

Nonlinear nummerical methods

`quad(fun,a,b)`	simpson integration of @fun from a to b
`fminsearch(fun,x0)`	minimum of unconstrained multivariable function using derivative-free method
`fmincon`	minimum of constrained function

Example: Constrained log-likelihood maximization, note the -

```
Parms_est = fmincon(@(Parms) -flogL(Parms,x1,x2,x3,y)
,InitialGuess,[],[],[],[],LwrBound,UprBound,[]);
```

Debbuging etc.

`keyboard`	Pauses exceution
`return`	resumes exceution
`tic`	starts timer
`toc`	stops timer
`profile on`	starts profiler
`profile viewer`	Lets you see profiler output
`try/catch`	Great for finding where errors occur
`dbstop if error`	stops at first error inside try/catch block
`dbclear`	clears breakpoints
`dbcont`	resume execution
`lasterr`	Last error message
`lastwarn`	Last warning message
`break`	Terminates executiion of for/while loop
`waitbar`	Waiting bar

Data import/export

`xlsread/xlswrite`	Spreadsheets (.xls,.xlsm)
`readtable/writetable`	Spreadsheets (.xls,.xlsm)
`dlmread/dlmwrite`	text files (txt,csv)
`load/save -ascii`	text files (txt,csv)
`load/save`	matlab files (.m)
`imread/imwrite`	Image files

Programming commands

`return`	Return to invoking function
`exist(x)`	checks if x exists
`G=gpuArray(x)`	Convert varibles to GPU array

```
function [y1,...,yN] = myfun(x1,...,xM)
```

Anonymous functions not stored in main programme

```
myfun = @(x1,x2) x1+x2;
```

or even using

```
myfun2 = @myfun(x) myfun(x3,2)
```

Conditionals and loops

```
for i=1:n
   procedure      Iterates over procedure
end               incrementing i from 1 to n by 1

while(criteria)
   procedure      Iterates over procedure
end               as long as criteria is true(1)
```

```
if(criteria 1)              if criteria 1 is true do procedure 1
   procedure1
elseif(criteria 2)          ,else if criteria 2 is true do procedure 2
   procedure2
else                        , else do procedure 3
   procedure3
end

switch switch_expression    if case n holds,
case 1                      run procedure n. If none holds
   procedure 1              run procedure 3
case 2                      (if specified)
   procedure 2
otherwise
   procedure 3
end
```

General comments

- Monte-Carlo: If sample sizes are increasing generate largest size first in a vector and use increasingly larger portions for calculations. Saves time+memory.
- Trick: Program that (1) takes a long time to run and (2) doesnt use all of the CPU/memory ? - split it into more programs and run using different workers (instances).
- Matlab is a column vector based language, load memory columnwise first always. For faster code also prealocate memory for variables, Matlab requires contiguous memory usage!. Matlab uses copy-on-write, so passing pointers (adresses) to a function will not speed it up. Change variable class to potentially save memory (Ram) using: int8, int16, int32, int64, double, char, logical, single
- You can turn the standard (mostly) Just-In-Time compilation off using: feature accel off. You can use compiled (c,c++,fortran) functions using MEX functions.
- Avoid global variables, they user-error prone and compilers cant optimize them well.
- Functions defined in a .m file is only available there. Preface function names with initials to avoid clashes, eg. MrP_function1.
- Graphic cards(GPU)'s have many (small) cores. If (1) program is computationally intensive (not spending much time transfering data) and (2) massively parallel, so computations can be independent. Consider using the GPU!
- Using multiple cores (parallel computing) is often easy to implement, just use parfor instead of for loops.
- Warnings: empty matrices are NOT overwritten ($[]+1=[]$). Rows/columns are added without warning if you write in a nonexistent row/column. Good practise: Use 3i rather than 3*i for imaginary number calculations, because i might have been overwritten by earlier. 1/0 returns inf, not NaN. Dont use == for comparing doubles, they are floating point precision for example: $0.01 == (1-0.99) = 0$.

Bibliography

Brown R, Hon FRS, MRSE, Acad RI, VPLS (1828) XXVII. A brief account of microscopical observations made in the months of June, July and August 1827, on the particles contained in the pollen of plants; and on the general existence of active molecules in organic and inorganic bodies. Philos Mag 4(21): 161–173. https://doi.org/10.1080/14786442808674769

Clauset A, Shalizi CR, Newman MEJ (2009) Power-law distributions in empirical data. SIAM Rev 51(4): 661–703

Michalet X (2010) Mean square displacement analysis of single-particle trajectories with localization error: Brownian motion in an isotropic medium. Phys Rev E Stat Nonlin Soft Matter Phys

Nørrelykke SF, Flyvbjerg H (2011) Harmonic oscillator in heat bath: exact simulation of time-lapse-recorded data and exact analytical benchmark statistics. Phys Rev E Stat Nonlin Soft Matter Phys 83(4):041103

Qian H, Sheetz MP, Elson EL (1991) Single particle tracking. Analysis of diffusion and flow in two-dimensional systems. Biophys J 60(4):910–921

Vestergaard C, Blainey PC, Flyvbjerg H (2014) Optimal estimation of diffusion coefficients from single-particle trajectories. Phys Rev E Stat Nonlin Soft Matter Physics

Wang YM, Flyvbjerg H, Cox EC, Austin RH (2007) When is a distribution not a distribution, and why would you care: single-molecule measurements of repressor protein 1-D diffusion on DNA. In: Controlled nanoscale motion: nobel symposium, vol 131, pp 217–240. Springer, Berlin/Heidelberg. ISBN: 978-3-540-49522-2. https://doi.org/10.1007/3-540-49522-3_11

Resolving the Process of Clathrin Mediated Endocytosis Using Correlative Light and Electron Microscopy (CLEM)

Martin Schorb and Perrine Paul-Gilloteaux

K. Miura, N. Sladoje (eds.), *Bioimage Data Analysis Workflows*, Learning Materials in Biosciences,
https://doi.org/10.1007/978-3-030-22386-1_6

What You Learn from This Chapter

This chapter will present the computational approach of registering images from different modalities based on manual selection of matching pairs of landmarks. Here we will present an image registration workflow based on MATLAB's image processing toolbox using the identification of sites of clathrin-mediated endocytosis by correlative light electron microscopy (CLEM) as an example. In the Appendix section, we will discuss the concept of image transformations and how to generate them based on pairs of landmarks. We will also learn how to fit a 2D Gaussian for a more accurate positioning of the landmarks.

6.1 Introduction

The purpose is to use Fluorescence Microscopy (FM) to localize clathrin vesicles, and to correlate it with Electron Microscopy (EM) to identify their ultrastructure (Fig. 6.1), based on the use of beads, as it was done in Avinoam et al. (2015). We will introduce the basic concept of image registration and dedicated MATLAB image processing commands to register light microscopy images of clathrin-mediated endocytosis and corresponding electron microscopy images to reveal the underlying ultrastructure (Avinoam et al. 2015). We will also discuss how enhancing the localization accuracy of fluorescence signals will improve the registration accuracy.

The first task in a typical CLEM experiment is to identify the two image datasets to be registered. The data from the second image modality (in the CLEM case this is EM) will likely be acquired in a targeted approach using the previous light microscopy observations. For a good review about different approaches of targeting the same area, the reader can be referred to de Boer et al. (2015).

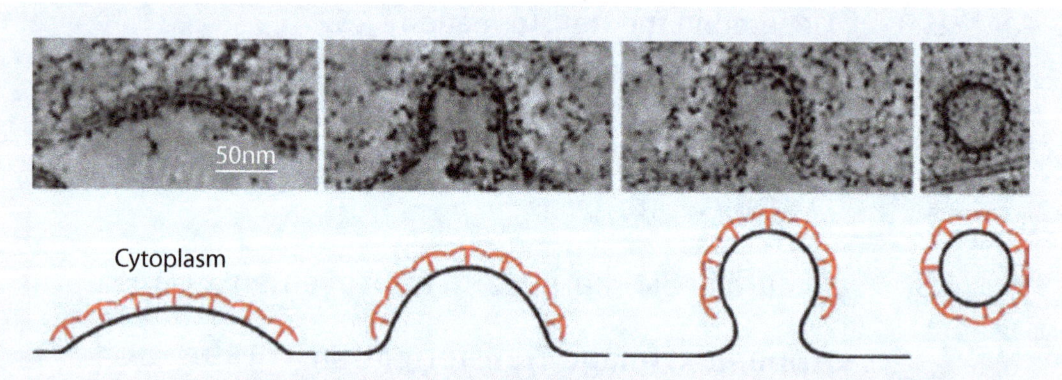

Fig. 6.1 Clathrin in conjunction with other proteins involved in endocytosis forms a lattice that can dramatically change the shape of the plasma membrane to form a vesicle. Top row: electron microscopy (EM) image. Bottom row: schematic of top row, with the plasma membrane in black and clathrin and associated proteins in red. (Image provided by Ori Avinoam, EMBL and Weizmann Institute of science, from the data published in Avinoam et al. 2015)

6.2 Data Presentation

All data used are available using the DOI:
▶ http://doi.org/10.5281/zenodo.1473535

The data we will use here were acquired using the protocol described in Avinoam et al. (2015). In order to obtain a higher-resolution insight into the ultrastructure underlying a fluorescence signal, we acquired images at high magnification (pixel size ≈ 1 nm, FOV 2 μm). The field of view at this magnification however is too small to contain enough landmark beads for a direct registration of the FM data. Therefore we need to first register to a lower magnification EM overview and then to the final high-resolution EM data (see ◘ Figs. 6.2 and 6.6 for an overview of datasets and scales).

Important note: on EM images the polystyrene beads appear as extended gray circles, not as black spots. The small black spots spread across the EM image are gold particles used to register EM data on itself for tilt correction (tomographic reconstruction) or alignment. We will use these gold beads as landmarks to accurately register the EM images of different magnifications.

The MATLAB functions described here will all handle 2D image data. Therefore we need to reduce the EM source data from the tomographic volumes. We can either choose single slices or an average of a small subset (5–10) slices from the source volumes. This can easily be done in Fiji (Schindelin et al. 2012), Icy (de Chaumont et al. 2012) or similar software. Here we could also adjust the contrast/brightness of the images to facilitate recognition of landmark features later in the process.

We already have prepared the matching images from FM and EM and made them available in the data directory. Ideally, we could now run an automated routine that would provide us with appropriate landmark features in both images. However, as the image data from FM and EM are intrinsically complementary and thus very different, in most of the cases the landmarks need to be selected manually.

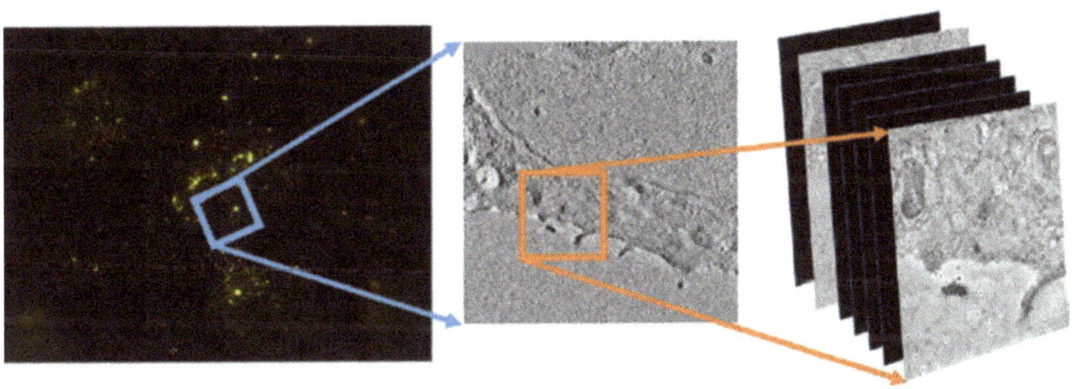

◘ **Fig. 6.2** We will register the fluorescent image ex1_FM.tif on the EM data ex1_EM.tif using beads visible in both modalities. We will then register ex1_EM.tif on ex1_highmag.tif which is acquired at the same depth of the sample using black gold particles. The position of interest showing clathrin is deeper in the cell, and appears as the other non black slice in this figure in the high mag EM stack. It is named ex1_highmag_poi.tif in this chapter

We can find the respective preprocessed files for the 2D registration work flow in the data directory:

- FM data: multi-color wide-field fluorescence acquisition in these channels: (red: RFP, green: GFP, blue: 380 nm emission for beads) `ex1_FM.tif`
- EM data low-magnification: we usually acquire a tomographic tilt series of images that after reconstruction results in a three-dimensional image stack. As the beads are located on top of the specimen, they can be identified in this stack. Usually we use an average of about 5 slices to have the best signature of the beads for the 2D registration. `ex1_EM.tif`
- EM data high-magnification: this comes from a high-resolution tomographic stack. For the 2D registration, we use a single slice that resembles the section selected for the low-mag registration, at the top of the tomogramm where the beads are. `ex1_highmag.tif`
- EM data high-magnification point of interest: this is a single slice from the same tomogramm that contains the structure of interest (here potential clathrin vesicles). `ex1_highmag_poi.tif`

An example of expected result navigating between these scales and images is shown on Fig. 6.9, where the red fluorescent spots are associated with specific ultrastructures.

6.3 Overview of Data Processing

- Step 1: Read and display EM and FM images
- Step 2: Manually identify landmarks pairs
- Step 3: Refine localisation by Gaussian fitting of the landmarks beads localization
- Step 4: Compute the rigid transform + scale from the list of paired landmarks localization.
- Step 5: Apply the Transformation (and discuss interpolation and potential artifacts)
- Step 6: Evaluate the confidence in structure matching

6.4 Tools Description

All codes are available here:
▶ http://doi.org/10.5281/zenodo.1473535

We will use the commercial software package MATLAB as well as its Image Processing and Optimization toolboxes (alternatively to the optimization toolbox, we can use the CurveFitting toobox; both versions of the code are provided). MATLAB provides a set of registration tools, gathered under the topic "image registration" in the MATLAB documentation (mat).

In particular, we will use:

- `cpselect`: built-in function from MATLAB Image Processing Toolbox allowing to provide a user interface for the selection of landmarks pairs.These are called control points in MATLAB language.
- `fitgeotrans`: built-in function from MATLAB Image processing toolbox allowing to fit a defined geometric transform matching pairs of landmarks (control points in Matlab language).

To refine the localisation of fluorescent spots correponding to the landmark beads, we will fit a 2D Gaussian to the beads' signal in a cropped image (see Fig. 6.11). For this procedure, there are two possible toolboxes in MATLAB: the Curve Fitting Toolbox, or the Optimization Toolbox. In particular, we will use:

- if we use the curve fitting toolbox we can use `fit`: built-in function from MATLAB Curve Fitting Toolbox allowing to find the parameters of a function which best fits given data. The advantage is that it provides also confidence intervals in fitting.
- if we use the optimization toolbox we can use `lsqnonlin`: built-in function from MATLAB Optimization Toolbox allowing to find the parameters of a function which best fits given data by least square fitting.

In our case, the function will be a 2D Gaussian equation and the data will be the pixels values of the cropped image around the manually selected landmarks.

As usual with Matlab, after downloading the code and data, we need to update our matlab path to include the code directory. The code directory contains files with the cascade of code used, as a correction or catch-up hint. It also contains the 2D Gaussian fitting functions.

6.5 Application to a CLEM Experiment

6.5.1 CLEM Workflow Overview and Preparation

The procedure of registering light microscopy data to electron microscopy data requires the landmarks to be clearly visible in both imaging modalities. We found fluorescently labelled polystyrene beads to match these criteria best (Kukulski et al. 2011). We do not want their signal to interfere with the fluorescence signal of interest, therefore the beads need to fluoresce in another channel. When using beads that only fluoresce in a different channel, shifts between the channels due to optical aberrations or stage instabilities during the acquisition will deteriorate the registration accuracy of our signal of interest. We either need to correct for these shifts or choose fluorescence beads that are both visible in the channel of interest and in another channel in order to be able to distinguish them from the real feature we want to localize. This is the case for our test dataset (Avinoam et al. 2015). The beads will be our landmarks (or control points in Matlab language). The typical feature of interest, an intracellular structural or morphological feature would have a size of about 100 nm. This is way beyond the diffraction limit of conventional fluorescence microscopy. The typical pixel size of FM data is on the order of 80–100 nm, so a single pixel difference in localization in the FM image could distort the registration of our feature by 100%. Therefore it is necessary to perform a precise sub-pixel localization not only on the fluorescent signal of interest but also on those of the landmarks.

The workflow to register the light microscopy data onto the EM data will be the following:

- identify the area of interest based on the target fluorescence signal (step 1)
- identify the locations of the surrounding fluorescent beads using the different channels
- identify the location of the beads visible in the EM image
- mark the matching landmark pairs in both images (step 2)
- precisely determine their localization in the FM image (step 3)
- calculate the image transformation (step 4)
- evaluate the confidence in structure matching (step 5)

- mark the coordinate of the feature of interest
- precisely determine its localization in the FM image
- apply the coordinate transformation
- create the output data (image overlays, coordinate lists, …)

The field of view in which we observe the EM features is not sufficiently large to capture enough landmark beads. Therefore, we need to perform an initial registration of the FM data with a lower-magnification overview and then a second registration to the higher magnification EM data. This second registration can be done in an automated fashion, as it is basically just a change in magnification and thus the image features are the same.

6.5.2 Labeling of Landmark Pairs

MATLAB's Image Processing Toolbox offers a selection of graphical tools to mark positions in an image. The command to mark coordinates in a displayed image is be `ginput`. However, we would like to assign coordinate pairs in the two image modes simultaneously. Therefore, the tool of choice is `cpselect`. This function expects the two images to be displayed as input and will give us two coordinate lists as a result of the point selection. The first image and associated coordinates are referred-to as "moving" and the second as "fixed". This means that the first image is the one whose coordinate system will be transferred onto the second. Note: Moving image is also sometimes called source image, and fixed image called target image.

6.5.2.1 Correlation from Low Magnification Tomogram to High Magnification EM Image

In order to locate the feature of interest in or high-resolution dataset, we need to register the low-magnification EM data, where we will map the FM data onto, to the high-magnification images. We can do this using the exact same tools as for registering the FM data with the low-magnification EM data. We will do this procedure first to get familiar with the tools, as we have very similar features in both images. We will use the gold beads (black spots) present on the specimen as common landmarks to register the EM images of different resolutions (Fig. 6.3).

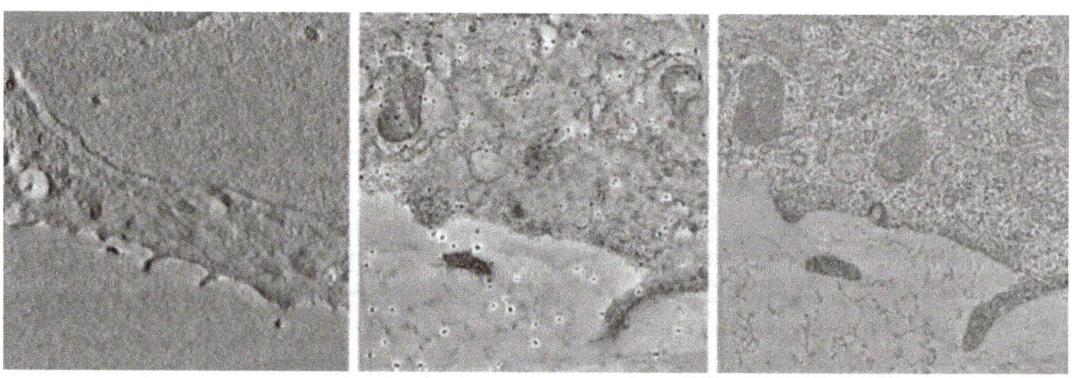

Fig. 6.3 The different images used in the registration of low-magnification to high-magnification EM data. Left: low-mag image (`em`), middle: high-mag image at the same z-height containing the gold beads as landmarks (`hm`), right: the slice of interest (`sm`)

First, we want to load our images. em is the overview (low-magnification) EM image, hm is the high-magnification image at the same z-height that also contains the gold beads we will use as landmarks. The slice of interest that contains the ultrastructure is also loaded (sm).

```
hm = imread('ex1_highmag_1.tif');
sm = imread('ex1_highmag_poi.tif');
em = imread('ex1_EM.tif');
```

We can display either of the images using the imshow command.

```
imshow(sm);
```

Now, we can open the two images next to each other in cpselect. The coordinates of landmark beads from the low-magnification image will be stored in c_lm, and those from the high-magnification EM image in c_hm. We have to make sure that we always click the corresponding landmark pairs in an alternating fashion between the two images to keep their correct association. The UI offers the option to use the already determined pairs (>2) for a prediction of the corresponding point for each new clicked spot. Simply activate the second "Add point and Predict Match" toggle button on the top left of the images (Fig. 6.4). Because we would like to store the clicked coordinates in two variables

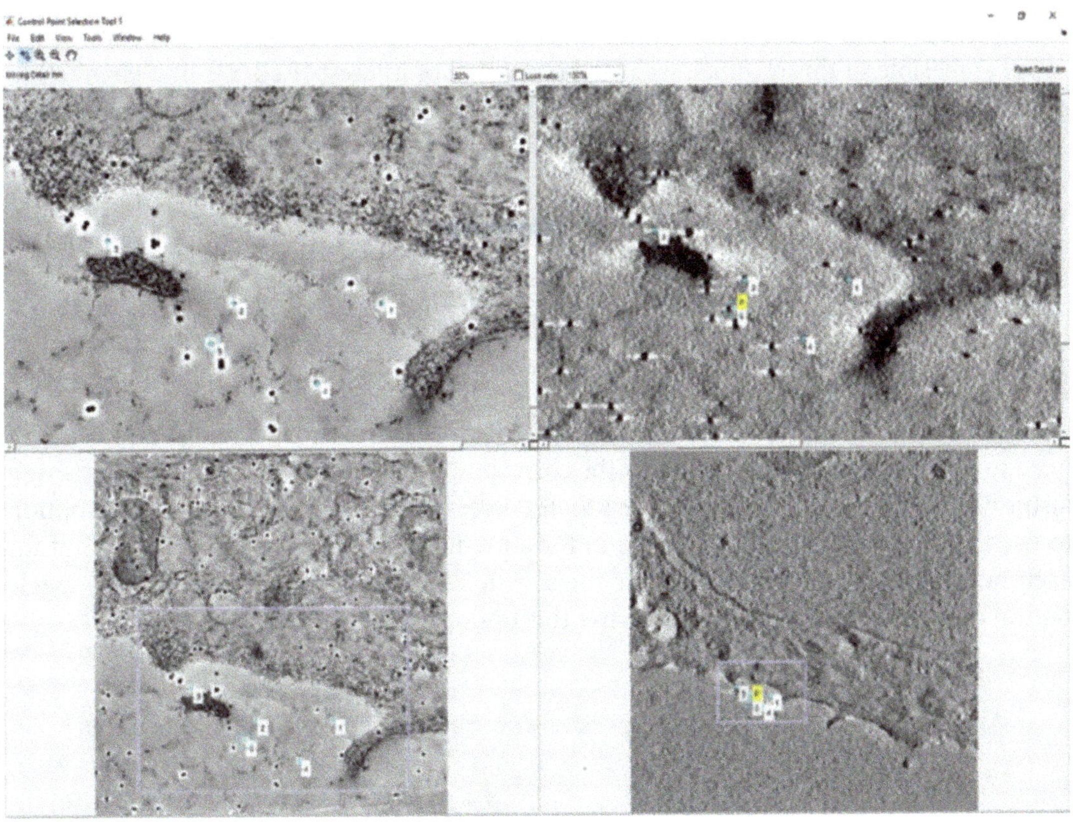

Fig. 6.4 The manual selection process of landmark pairs in MATLAB's cpselect tool showing EM images at different magnifications. The "Add point and Predict Match" tool is selected and the predicted position for point 4 is indicated in the right panel in yellow

directly, we need to provide the `'Wait',true` option to `cpselect` to prevent MATLAB from running other processes while we pick the coordinates.

```
[c_hm,c_lm] = cpselect(hm,em,'Wait',true);
```

Once done, simply close the `cpselect` window. The two output variables `c_hm` and `c_lm` contain the two coordinate lists of the landmark pairs that we need to generate our transformation.

We will generate the transformation that best relates these two sets of coordinates in the ▶ Sect. 6.5.5.

Now that we have seen the `cpselect` tool, we can use it to defind the landmark pairs for aligning the light-microscopy data to the low-magnification EM data.

First, we want to load our images. `fm` is the fluorescence image (FM), `em` will be the overview image from EM.

```
fm = imread('ex1_FM.tif');
```

We can display either of the images using the `imshow` command and automatically adjust the image contrast with `imadjust`. The FM image is a 16 bit 3-channels image. In order to work with it using the described Matlab tools, we need to choose a single channel to work with. In order to avoid chromatic aberrations in the registration process, and as the beads are visible in this channel, we use the channel of interest (Red, i.e. first channel) to mark the beads. Note that the blue channel (channel 3) contains only the beads, and could be used later one to differentiate beads and Clathrin.

```
fm1 = fm(:,:,1);   % select red channel
imshow(imadjust(fm1));
```

Open the two images next to each other in `cpselect`. The coordinates of landmark beads from the FM image will be stored in `c_fm`, those from the EM image in `c_em`. We need to make sure that we always click the corresponding landmark pairs in an alternating fashion between the two images to keep their correct association. The UI offers the option to use the already determined pairs (>2) for a prediction of the corresponding point for each new clicked spot. Simply activate the second "Add point and Predict Match" toggle button on the top left of the images to use this option.

```
[c_fm,c_em] = cpselect(em,imadjust(fm1),'Wait',true);
```

Because the polystyrene beads we use as landmark markers are very difficult to identify on the EM data (position of beads in EM data are shown in Fig. 6.10, some initialization points are provided for the ease of this demonstration. Let's load their coordinates (this file is in the code directory):

```
load('preselectedpoints.mat');
```

This mat-file contains two variables:
em_cp_preselected and fm_cp_preselected.

```
[c_fm,c_em] = cpselect(em,imadjust(fm1),em_cp_preselected,fm_cp_
preselected,'Wait',true);
```

The cpselect window should now look like on Fig. 6.5.

Try to add a point using the "Add point and Predict Match" toggle button.

Once we are done with selecting landmark pairs, close the cpselect window. The two output variables c_fm and c_em contain the two coordinate lists, in pixels, that we need in order to generate our transformation.

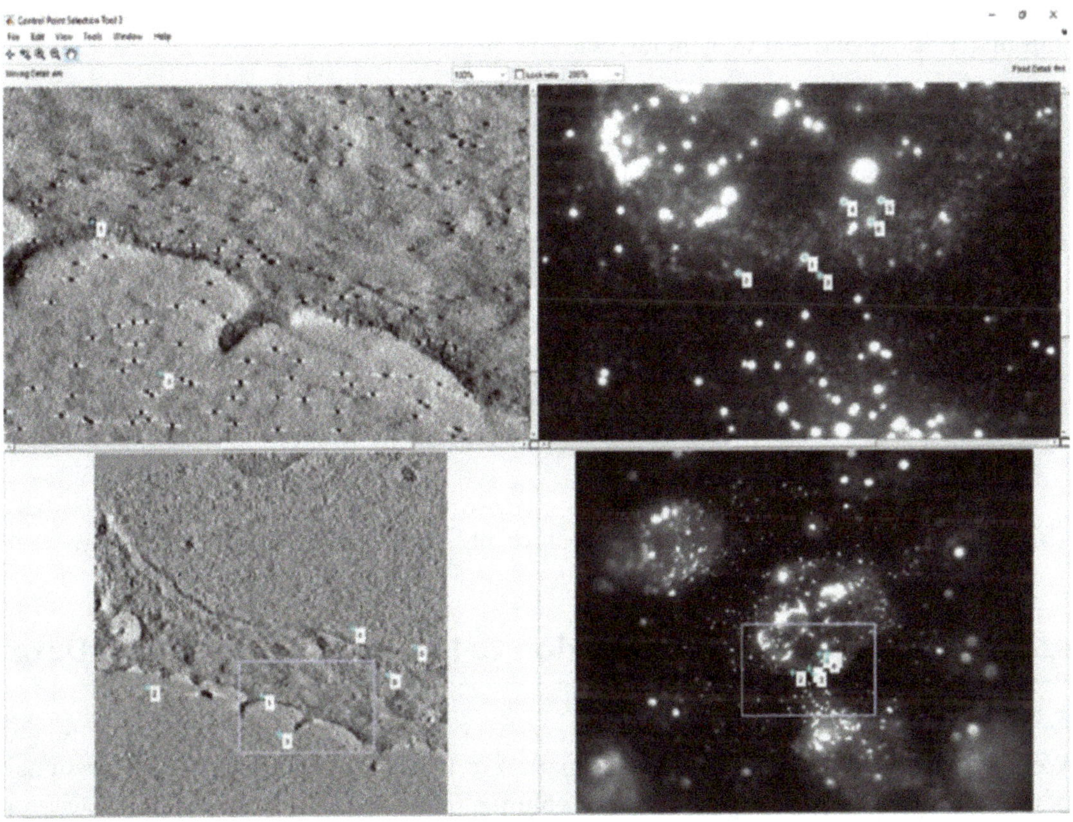

Fig. 6.5 The manual selection process of landmark pairs in light (right) and electron microscopy (left) images using MATLAB's cpselect tool

6.5.3 Generating the Transformation

In the case of a similarity, we would like to solve for the transformation matrix T in the system of equations described in equation 6.14. The lists of coordinates now consist of row vectors instead of the required column vectors, so we need to transpose them before the calculation:

```
T = c_em' / [c_fm';ones(1,length(c_fm))];
```

with the result giving us a transformation matrix like this:


```
T =

    1.0e+03 *

      0.0125    -0.0038    -5.2820
      0.0040     0.0125    -9.1019
```

If we compare the (1,2) and (2,1) entries of the matrix, i.e. 0.0040 and −0.0038 (the coefficient b from Eq. 6.10), we notice that the solution does not exactly fulfill the prerequisites of a similarity (same magnitude in scaling in both axes). Let's check what the transformation matrix looks like that we generate with `fitgeotrans`:

```
structT = fitgeotrans(c_fm,c_em,'similarity');
T = structT.T';

T =

    1.0e+03 *

      0.0124    -0.0040    -5.1138
      0.0040     0.0124    -9.0235
           0          0     0.0010
```

This matrix resembles the one generated before, but now describes a true similarity.

6.5.4 Applying the Transformation to Image and Coordinate Data

6.5.4.1 Transforming Images

We now would like to apply the transformation to find out where our fluorescent signal of interest is located within the EM image/volume. In order to transform an image we can use the MATLAB function `imwarp`. Obviously our initial FM image covers a much larger field of view than the EM image (Fig. 6.6). We therefore need to provide the function with the scale and dimension of the target image. This is done using `imref2D`.

In order to generate the pair of registered images, we will now apply the transformation in the structured variable `strucT` computed by `fitgeotrans`, using `imwarp`.

```
em_geom = imref2d(size(em));
fm_trans = imwarp(fm1,structT,'OutputView',em_geom);
figure(1);imshowpair(em,fm_trans,'montage');
 %shows the images side-by-side
figure(2); imshowpair(em,fm_trans,'blend');
 %shows them merged
```

The result of these commands is shown in Fig. 6.7 (for one channel only).

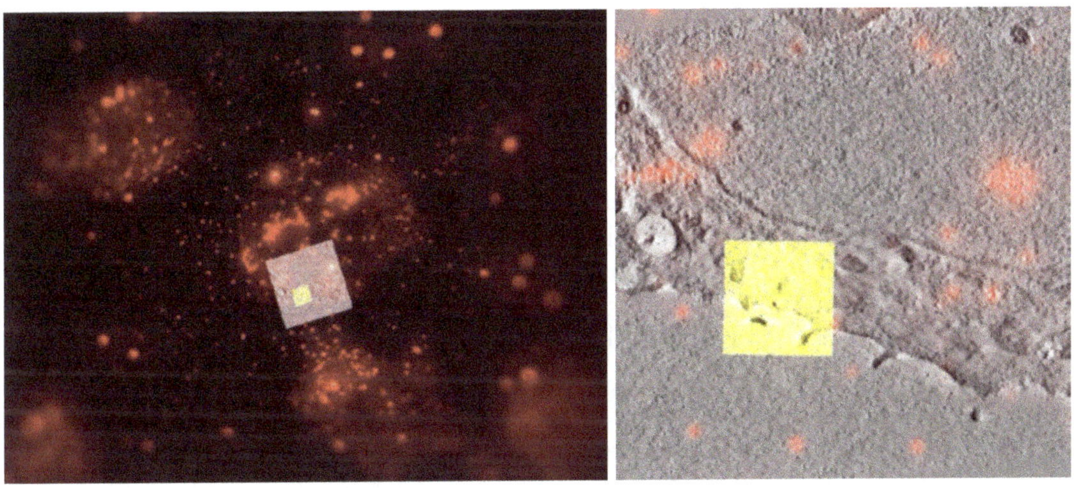

Fig. 6.6 Illustration of the relative size of EM and FM images. The EM images (low mag and high mag, size: 2048 × 2048 pixel) are overlaid for comparison of scale, after registration. Red: red channel (showing clathrin and beads) from FM, Gray: low mag EM. Yellow: High Mag EM after registration on the low mag EM

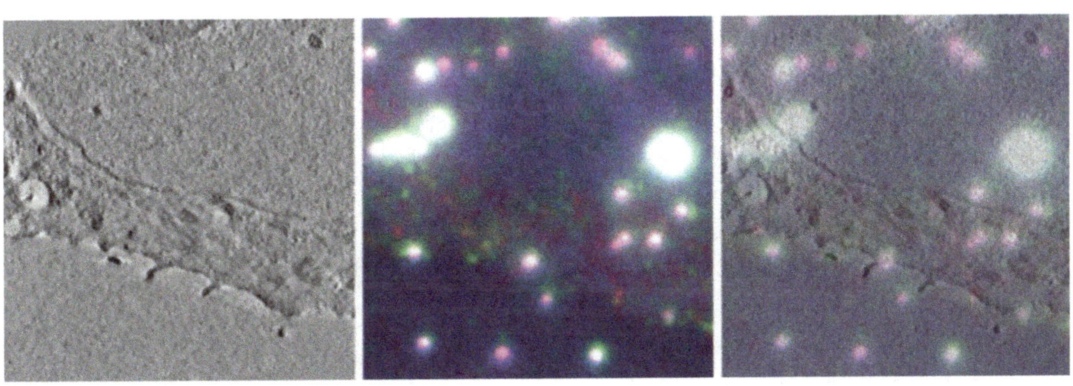

Fig. 6.7 The final result of the image registration (left and center: 'montage' parameter) together with the overlay using the 'blend' option (right) for `imshowpair`

6.5.4.2 Transforming Coordinates

In order to accurately transfer a set of coordinates we can use `ginput` on the fluorescence image to select points of interest on the FM image. Several points can be entered. Do not forget to press the return key when done to quit ginput mode. We then transform the obtained coordinate list using `transformPointsForward` and display the resulting coordinates on top of the EM image.

```
figure(3);
imshow(imadjust(fm1));
[x,y] = ginput;
[u,v] = transformPointsForward(structT,x,y);
figure(2), hold on, plot(u,v, '*r');
```


6.5.5 Registering the Low-Magnification and the High-Magnification EM Data

Now that we have transformed the pixel coordinates of the feature(s) of interest from the FM image onto the low-magnification EM image, we need to perform a second transformation in order to map these coordinates to the high-magnification data. Let's use the landmark lists that we have generated in the beginning (of ▶ Sect. 6.5.2) and calculate the second registration between low and high EM magnifications.

```
lm2hm = fitgeotrans(c_lm,c_hm,'similarity');
```

In order to test whether the transformation is correct, we would like to display the high-mag image in the context of the lower-mag overview. Therefore we need to warp it using the inverse transform.

```
hm2lm = invert(lm2hm);
hm_trans = imwarp(hm,hm2lm,'OutputView',em_geom);
imshowpair(em,hm_trans);
```

The function `imshowpair` without options will display the result of the registration in a color overlay (◘ Fig. 6.8).

Let's now apply the transformation to the high-magnification data to our target coordinates of the features of interest. We have already found their coordinates in the low-mag EM (`u,v`), so we need to apply our transformation `lm2hm` to these.

```
[x_final,y_final] = transformPointsForward(lm2hm,u,v);
figure;imshow(sm);hold all
scatter(x_final,y_final,100, 'go');
%this code generates the lowest panel in the results figure
```

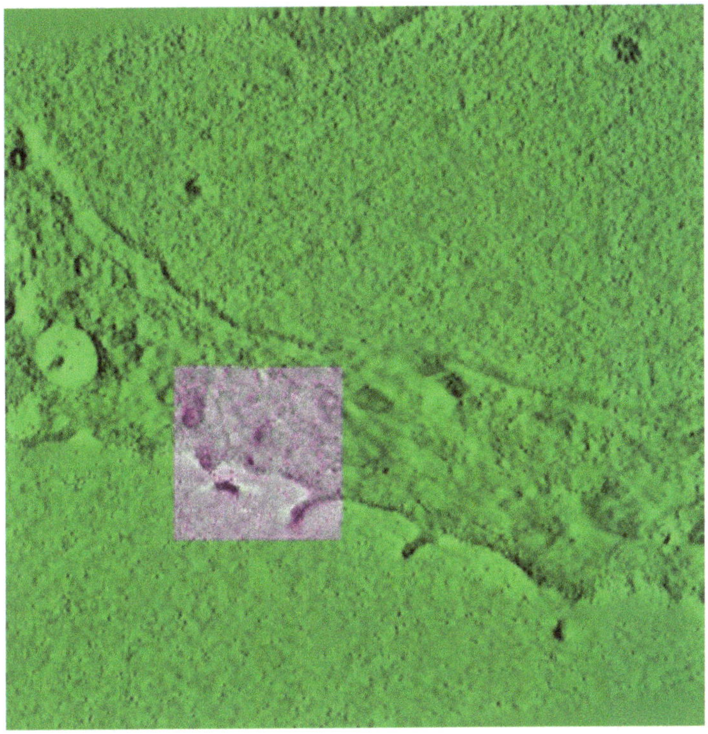

Fig. 6.8 The color overlay produced by imshowpair allows to check successful registration of the high-mag image to the low-mag

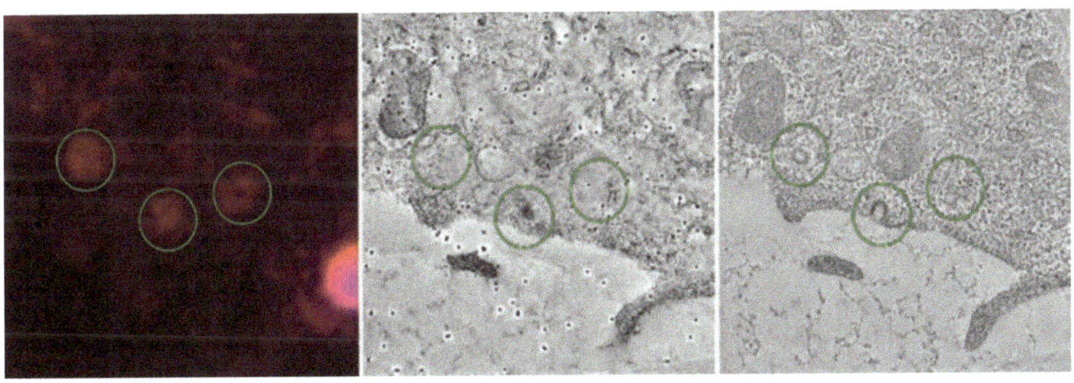

Fig. 6.9 Left: Clathrin signal identified by FM (green circles), note the bright signal of a polystyrene bead on the right; Middle: High mag EM top slice showing the gold beads and the polystyrene bead (right) on the surface of the specimen; Right: High mag EM slice some nm below in the tomogram showing the signal originating from complete vesicles (left, right) and from a forming invagination (center). This panel is generated by the example code

The scatter function here will draw green circles on top of the existing figure (high-mag EM) at the coordinates of the transformed positions (Fig. 6.9).

Exercise: Try to re-create the top panel of Fig. 6.9.

6.6 Accuracy Estimation and Improvements

With the transformation matrix we obtained, we can calculate the transformed coordinates of our landmarks from FM to EM and compare them with the clicked positions.

```
c_em1 = T * [c_fm';ones(1,length(c_fm))];
 % identical alternative: c_em1 = transfromPointsForward(a,c_fm);
figure; imshow(em);
hold all;
scatter(c_em(:,1),c_em(:,2),70, 'r+');
scatter(c_em1(1,:),c_em1(2,:),70, 'bo');
 % determine the deviation for the predicted points and get average and standard deviation
pos_diff = c_em1(1:2,:)' - c_em;
[mu,sig] = normfit(pos_diff);
```


This will give us an idea about the deviation of the landmark positions from their predictions in EM pixels (Fig. 6.10). Optionally, we can try different types of transformations and compare the resulting deviations.

It is advised to check what happens to the matrix when the control points are moved to slightly different positions (using `cpselect`).

In the example we skipped step 3, the refining of FM coordinates using a Gaussian fit. In order to improve the accuracy of the registration, both the localization of the fluorescence signal of interest and of the landmark beads can be improved using a fit of the peak with sub-pixel accuracy. This fit can be performed directly in MATLAB during the workflow. We are going to add this step, just after the manual selection, for the FM fluorescent signal of the beads (Step 6.5.2).

For this we can use the provided script called `GaussianFit_...` that corresponds to the Toolbox available on our computer (either optimization or curve fitting Matlab toolbox). We should start by removing the suffix of the file needed such that it is called `GaussianFit.m` alone. To know which toolbox is available, type `ver`. The principle is the following:

- Input parameters are the original FM image (`fm1`), the list of selected control points (or landmarks) on FM (`c_fm`), and a parameter in pixels that will give the crop size, called `N`.
- For each control point
  - Crop the original image around the control point position, plus and minus `N`.
  - Try to fit a 2D Gaussian (we are assuming that the Gaussian is symmetric, i.e $\sigma_x = \sigma_y = \sigma$)

$$G(x,y) = Ae^{-\left(\frac{(x-x_0)^2+(y-y_0)^2}{2\sigma^2}\right)} \tag{6.1}$$

  - Visually check the quality of the fit in a plot.
  - The parameters of this Gaussian will give us: A the amplitude of the Gaussian (peak height), σ its width, and (x_0, y_0) the central point of the Gaussian (peak position). Here we do not really care about the first two parameters, but (x_0, y_0) will allow us to correct the original position of the control point (x, y). Note that in a more advanced script, A and σ could be used for discriminating bad fitting automatically.

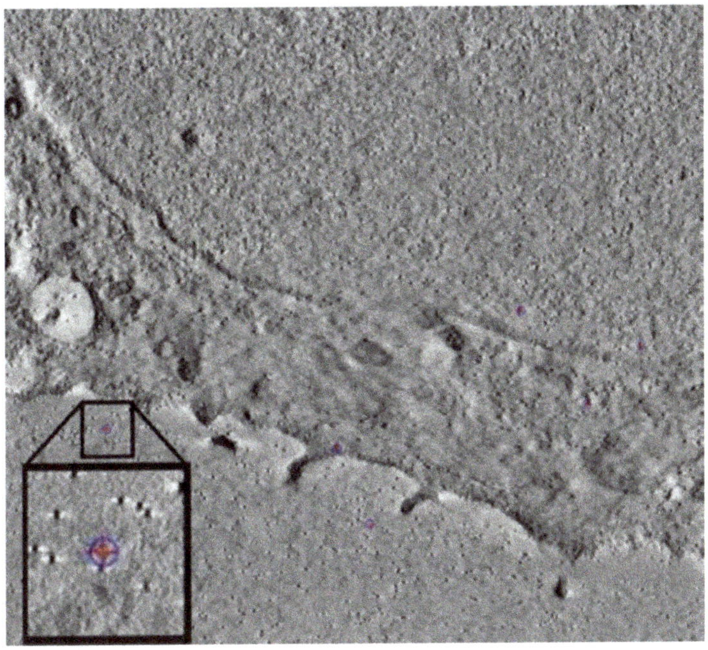

Fig. 6.10 Comparison of the coordinate predictions for the landmark points (blue circles) with their initial clicked positions (red crosses)

After having read the script, run it. $N=5$ pixels will create a crop area of 11×11, which should be sufficient in our case.

```
corrected_positions = GaussianFit(fm, c_fm, 5);
```

It should display the fitting and the original data as shown in Fig. 6.11. Press any key to process the next control point (this is achieved by a `pause` command in the MATLAB script).

Check the results visually with `cpselect`, now showing the corrected positions on top of the image.

```
[c_fm,c_em] = cpselect(imadjust(fm1),em,corrected_positions,c_em,
'Wait',true);
```

Exercise: Complete `SimpleCLEMworkfow.m` by placing the accuracy refinement by Gaussian fitting. Add also a plot of the relative position of FM control points to matching control points in EM: it can indicates a bias such as a small drift if it is not centered around 0. We can use `normfit` to study their distribution. One could also add an histogram of the distance by using `hist`. Solution is provided in `CLEMworkflowwithstep3andsimpleerrorstudy.m`.

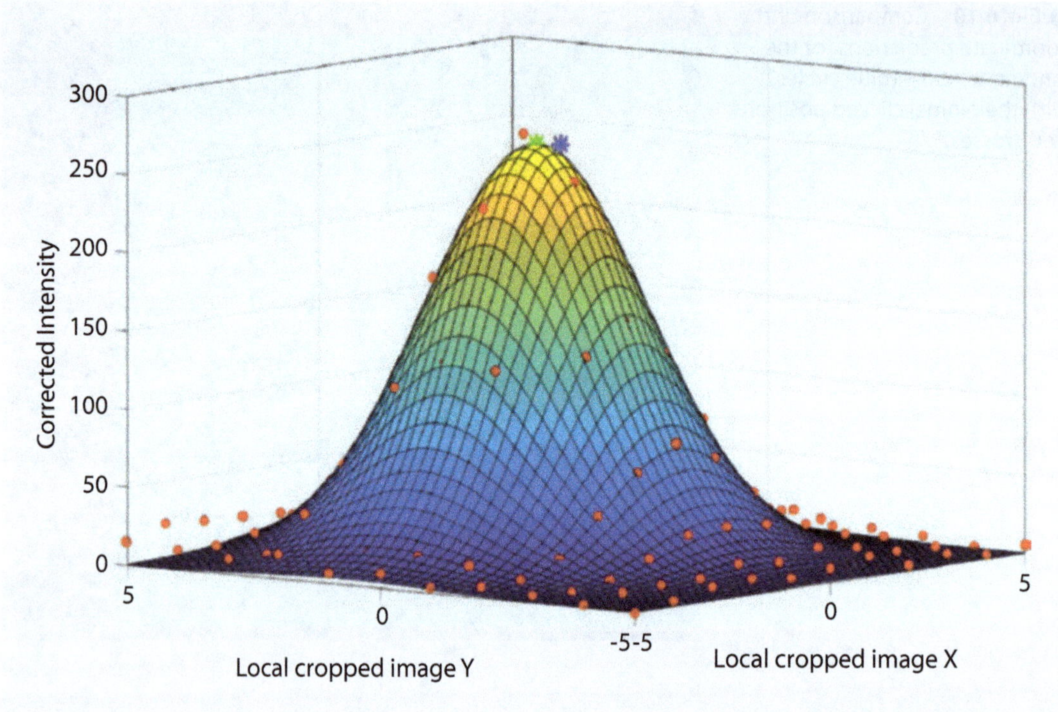

6

Fig. 6.11 Example of correction of manual selection based on Gaussian fitting. The mesh is the Gaussian fit, the red dots are the original pixel intensities in the cropped image (corrected by their base value), the blue star is the original clicked position, the green star is the corrected position. Note that here the clicked values were already very good, but remember that 1 pixel correction is about 100 nm correction, i.e. about 100 pixels in the high magnification EM image

Take Home Message

In this chapter we have shown the basic principle of post-acquisition coordinate registration. We have seen how to express linear transforms with matrices, and how to compute them from pair of coordinates. We have registered the FM image showing polystyrene beads to an EM image at a low magnification, with enough beads visible in the field of view. We have then registered the low magnification EM to an higher magnification EM. Transforms can then be combined by simply multiplying them to position the fluorescence image on the high magnification EM image. The code as described here was used for a number of CLEM studies in the recent years (Kukulski et al. 2012; Schellenberger et al. 2014; Avinoam et al. 2015; Hampoelz et al. 2016; Curwin et al. 2016), but accuracy estimation was computed using another more effective approach than in this module. In this chapter, the localization error was computed only for the control points. Investigating only the error for the predicted landmark coordinates (as in 6.6) leads to an underestimation of the error for the points of interest. These might be located distant to landmarks and therefore behave differently under the transformation.

The accuracy of a registration also strongly depends on the accuracy of the localization of the landmarks. The Gaussian fitting can help to to reduce the resulting error. When an accurate localisation is not possible, another way of reducing registration inaccuracy is to increase the number of landmark points, and to make sure that they surround the point of interest (see Paul-Gilloteaux et al. (2017) for a theoretical description of accuracy in registration).

The registration workflow demonstrated in this chapter is applied in 2D, but the 3D workflow can be constructed in a similar way by adding the z dimension to both the coordinates and the transformation matrix.

There are alternative tools that perform similar tasks include ec-clem (Paul-Gilloteaux et al. 2017), which also supports 3D registration and propose some automatic registration options, as well as an estimation of error in any points of the image. In addition, it can help selecting the type of transformation needed, in particular it can automatically detect if a elastic transformation is needed, e.g. if the sample underwent deformations due to the fixation process for example. As it was demonstrated in Paul-Gilloteaux et al. (2017), selecting an elastic transformation when not needed will actually augment the error in other points of the images than the landmarks (Fig. 6.12).

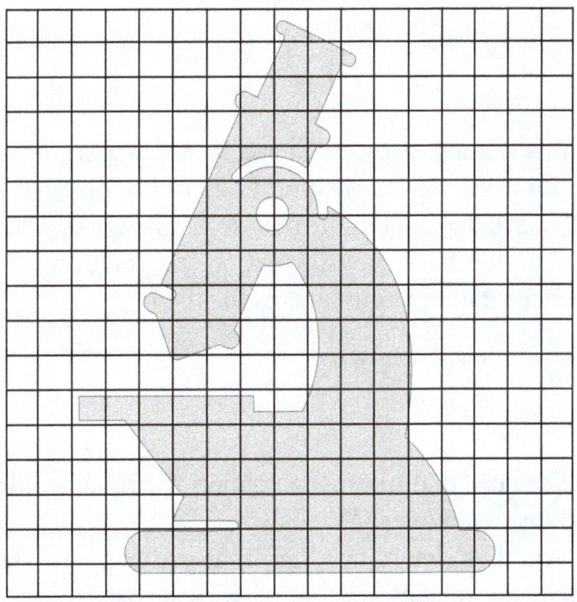

Fig. 6.12 The source image we will use to demonstrate the transformations. Microscope pictograph adapted from ant (2013)

Acknowledgements We thank Marion Louveaux (Heidelberg University) for reviewing this chapter.

Appendix: Image Transformations

Basic Similarity and Affine Transformations

The position of each pixel and each object inside an image is given by its two coordinates $\begin{pmatrix} x \\ y \end{pmatrix}$ to which an intensity value is associated.

Any linear transformation can be written as the multiplication by a matrix T that describes the transformation applied to the the image vectors.

$$\begin{pmatrix} x' \\ y' \end{pmatrix} = T * \begin{pmatrix} x \\ y \end{pmatrix} = \begin{pmatrix} a & b \\ c & d \end{pmatrix} * \begin{pmatrix} x \\ y \end{pmatrix} = \begin{pmatrix} ax + by \\ cx + dy \end{pmatrix} \tag{6.2}$$

Any linear transformation (i.e translations, rotations, scaling, etc…) can be seen as a combination of elementary transformations that can be represented as sequential matrix multiplications.

$$\begin{pmatrix} x' \\ y' \end{pmatrix} = T * U * V * \dots * \begin{pmatrix} x \\ y \end{pmatrix} \tag{6.3}$$

6

The simplest possible transformation is the uniform scaling with a constant s (Fig. 6.13). The transformation can then simply be described as the Identity matrix multiplied by this constant. This means, that in order to obtain this matrix from image data, we need to find one parameter (s).

$$\begin{pmatrix} x' \\ y' \end{pmatrix} = s * \begin{pmatrix} 1 & 0 \\ 0 & 1 \end{pmatrix} * \begin{pmatrix} x \\ y \end{pmatrix} = s * \begin{pmatrix} x \\ y \end{pmatrix} \tag{6.4}$$

In this scaling example, if we take a scaling of $s = 0.5$ (i.e. reducing the image size by 2), if a pixel was at position (2,2) in the original image it would then move to position (2*0.5,2*0.5)=(1,1) in the new scaled image (see left panel of Fig. 6.13)

Another basic transformation is a rotation (Fig. 6.13). Here the rotation matrix T, given a rotation angle θ, takes the form:

$$T = \begin{pmatrix} \cos\theta & -\sin\theta \\ \sin\theta & \cos\theta \end{pmatrix} \tag{6.5}$$

This means, that in order to obtain the rotation matrix from a pair of images, we need to find one parameter (θ).

Another elementary step that can happen when multi-modal images are compared is that the images are flipped with respect to each other. The transformation matrix to describe the swapping of the two coordinate axes looks like this:

$$T = \begin{pmatrix} 0 & 1 \\ 1 & 0 \end{pmatrix} \tag{6.6}$$

Why?

$$\begin{pmatrix} x' \\ y' \end{pmatrix} = T * \begin{pmatrix} x \\ y \end{pmatrix} = \begin{pmatrix} y \\ x \end{pmatrix} \tag{6.7}$$

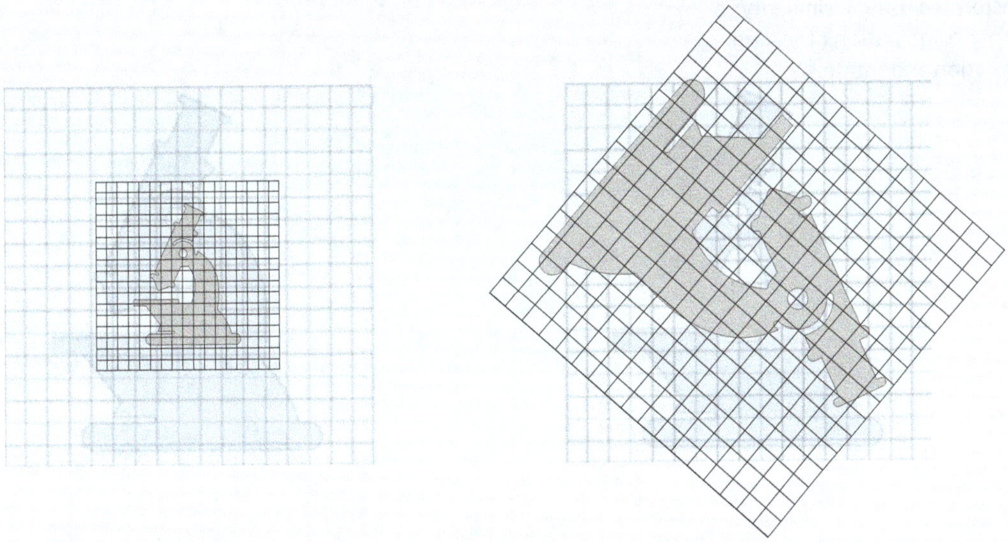

Fig. 6.13 The test image scaled by a factor of $s = 0.5$ (left) or rotated by 130° (right). The origin of the coordinate system is at the center of the image

Coordinates swapped!

Another elementary image transformation—the translation of coordinates—cannot be described using the aforementioned very elegant concept of nested matrix multiplications. So how can the following translation

$$\begin{pmatrix} x' \\ y' \end{pmatrix} = \begin{pmatrix} x \\ y \end{pmatrix} + \begin{pmatrix} t_x \\ t_y \end{pmatrix}$$

be written as a multiplication?

In order to represent a translation as a matrix multiplication we have to add an extra dimension to our description that does not correspond to a real coordinate dimension in our image data but only plays a role during the calculations. An image coordinate would then be written like this: $\begin{pmatrix} x \\ y \\ 1 \end{pmatrix}$. The matrix describing a translation is obtained by adding the column of the translation vector $\begin{pmatrix} t_x \\ t_y \end{pmatrix}$ to the two-dimensional Identity matrix.

$$\begin{pmatrix} x' \\ y' \end{pmatrix} = \begin{pmatrix} 1 & 0 & t_x \\ 0 & 1 & t_y \end{pmatrix} * \begin{pmatrix} x \\ y \\ 1 \end{pmatrix} = \begin{pmatrix} 1*x + 0*y + 1*t_x \\ 0*x + 1*y + 1*t_y \end{pmatrix} = \begin{pmatrix} x + t_x \\ y + t_y \end{pmatrix} \quad (6.8)$$

The combination of rotation and translation with an optional flip of coordinate axes is called **rigid transformation**. A rigid transformation conserves all geometrical properties of the original structure, such as areas and relative orientations.

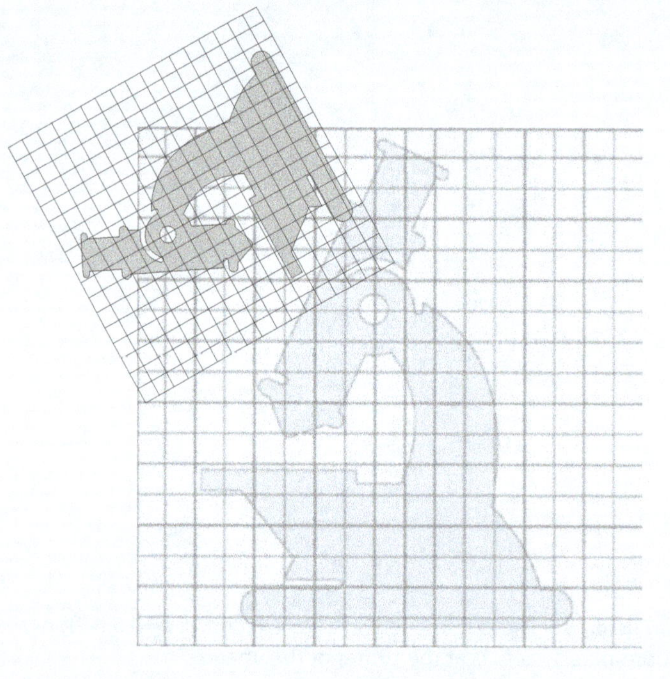

Fig. 6.14 The test image is transformed using a similarity: a linear combination of scaling, translation and rotation

When adding a uniform scaling, the transformation is called **similarity** (Fig. 6.14). Under a similarity, parallel lines remain parallel and angles are conserved. This implies that all shapes stay the same.

A general similarity can be written as:

$$\begin{pmatrix} x' \\ y' \end{pmatrix} = \begin{pmatrix} s*\cos\theta & -\sigma*s*\sin\theta & t_x \\ s*\sin\theta & \sigma*s*\cos\theta & t_y \end{pmatrix} * \begin{pmatrix} x \\ y \\ 1 \end{pmatrix} \tag{6.9}$$

In order to find a similarity matching two coordinate systems, the four unknown parameters s, θ, t_x and t_y have to be determined. The additional parameter σ which is ± 1 determines whether a coordinate flip is included or not.

When we allow a non-uniform scaling that affects the coordinate axes differently, the resulting transformation is called **affine** (Fig. 6.15) and no longer preserves angles and shapes, but parallel lines. The representation of an affine transformation requires to define all six matrix components.

$$\begin{pmatrix} x' \\ y' \end{pmatrix} = \begin{pmatrix} a & b & t_x \\ c & d & t_y \end{pmatrix} * \begin{pmatrix} x \\ y \\ 1 \end{pmatrix} \tag{6.10}$$

Higher-Order Transformations

When a coordinate frame is registered onto another coordinate system, the scaling factors that determine the stretching can be defined to vary in a linear fashion. Transformations

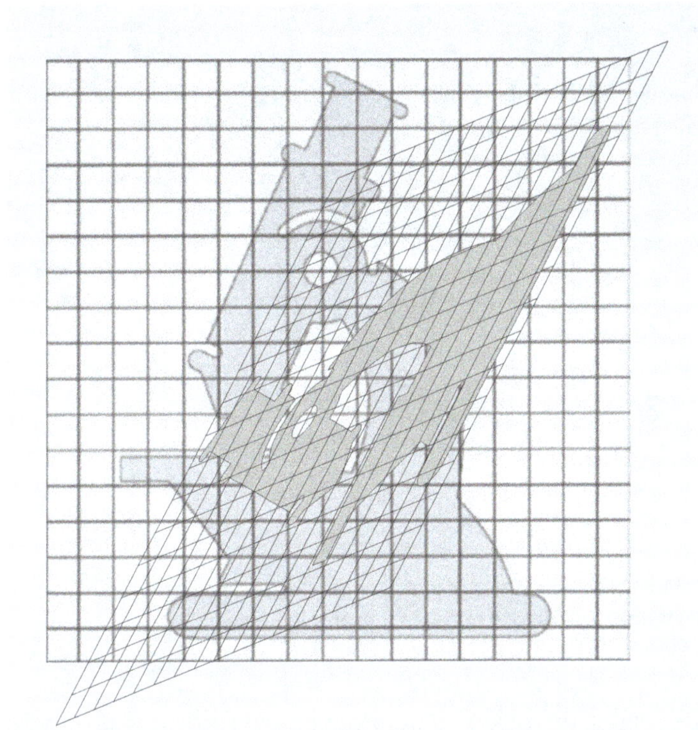

Fig. 6.15 The test image is transformed using an affine transformation. The stretch in one coordinate direction causes a distortion in shape

that add this flexibility to affine transformations are called **projections** (Fig. 6.16) and are described by a general 3×3 matrix with 9 unknown parameters. After a projection, straight lines will remain straight.

$$\begin{pmatrix} x' \\ y' \\ 1 \end{pmatrix} = \begin{pmatrix} a & b & c \\ d & e & f \\ g & h & i \end{pmatrix} * \begin{pmatrix} x \\ y \\ 1 \end{pmatrix} \quad (6.11)$$

Instead of applying a single matrix multiplication to the coordinates, an other way of mathematically describing such coordinate transformation would be to have the result depend on higher polynomial orders of the input.

For the second order

$$\begin{pmatrix} x' \\ y' \end{pmatrix} = T * \begin{pmatrix} x^2 \\ y^2 \\ x * y \\ x \\ y \\ 1 \end{pmatrix} \quad (6.12)$$

T is a 6×2 matrix with 12 unknown coefficients. With higher order polynomials or groups of transformations, where each only matches a local set of coordinates, any degree of flexibility can be achieved (Fig. 6.16). However, the higher the complexity of the approach,

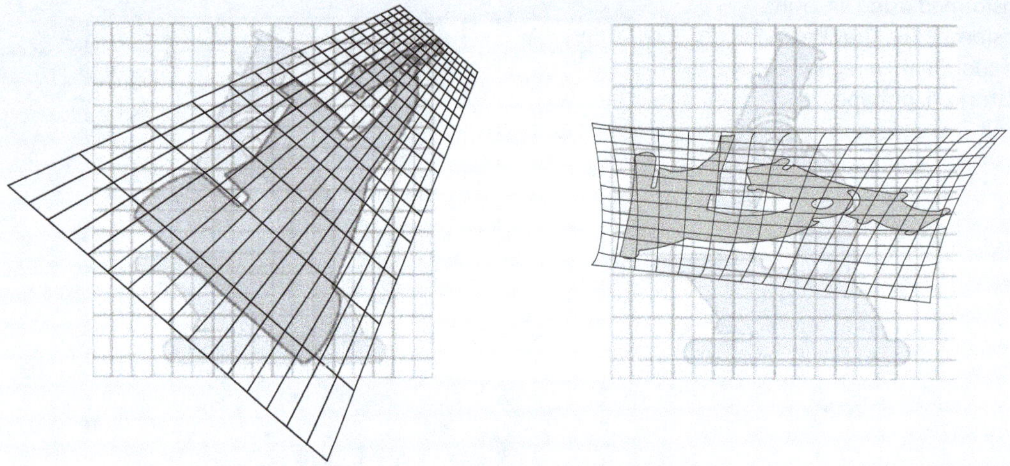

Fig. 6.16 Examples of a projection (left) and a non-linear image transformation (right). Note the loss of straight lines in the non-linear example image

the higher the risk of generating an overfitting that only represents the priors but not the true state of the entire system.

Generating Transformations from Image Coordinates

The goal of a post-acquisition correlative experiment is to localize of a feature from available image data inside a second, different image dataset that provides complementary information. In order to find the transformation that registers the coordinate frame from the first imaging modality to the second, we need to define the unknown parameters. In a typical CLEM experiment, the information from the light microscopy data and those obtained by EM are fundamentally different, so an automated feature detection will most likely fail due to the lack of common structures. We therefore rely on a generic approach and on the availability of landmark pairs that the user can manually position in both image modalities.

Let's assume we want to find the parameters of a similarity without reflection. We thus need to define the four coefficients for

$$\begin{pmatrix} x' \\ y' \end{pmatrix} = \begin{pmatrix} a & -b & t_x \\ b & a & t_y \end{pmatrix} * \begin{pmatrix} x \\ y \\ 1 \end{pmatrix} \tag{6.13}$$

with $a = s * \cos\theta$ and $b = s * \sin\theta$. This notation represents a set of two linear equations. Since each known coordinate pair will solve one set, we need three pairs of landmarks to solve all unknown pairs.

The higher the flexibility of the desired transformation, the higher the number of unknowns in the equations and therefore the more defined landmark pairs need to be provided.

The system of equations can thus be written as:

$$\begin{pmatrix} x'_1 & x'_2 & \cdots & x'_n \\ y'_1 & y'_2 & \cdots & y'_n \end{pmatrix} = \begin{pmatrix} a & -b & t_x \\ b & a & t_y \end{pmatrix} * \begin{pmatrix} x_1 & x_2 & \cdots & x_n \\ y_1 & y_2 & \cdots & y_n \\ 1 & 1 & \cdots & 1 \end{pmatrix} \tag{6.14}$$

or in MATLAB code

```
V = T * [U;ones(1,size(U,2))];
V = [x1, x2, x3; y1, y2, y3]; % x,y landmark coordinates in target
frame
```

with the matrices U containing the source and V the target coordinates.

```
U = [x1, x2 ,x3; y1, y2, y3];
```

In order to solve for the matrix coefficients in T, we simply need to invert this matrix multiplication. In general, matrix multiplications are not commutative, so the order of the factors matters. In our case, we want to determine the left factor, therefore we can apply MATLAB's "/" operator (identical to the function `mrdivide`).

```
T = V / [U;ones(1,size(U,2))];
```

However, this approach will solve the general matrix parameters and will not take into account the restrictions on our transformation matrix, such as the number of free parameters depending on the chosen type of transformation. It will simply determine a matrix that solves this set of equations. Moreover, if the number of provided landmark pairs exceeds the number of constraints necessary to solve the equations, MATLAB will find the matrix coefficients using a least-squares approach. In order to restrict the solution to a specific transformation type, we will use the MATLAB function `fitgeotrans`. It produces a MATLAB transformation structure that contains the matrix and some metadata. It requires the coordinate data in columns, so we transpose U and V.

```
Transformation_Type= 'nonreflectivesimilarity';
structT=fitgeotrans(U',V',Transformation_Type);
T=structT.T'
```

The script code for these 2 methods can be found in `Background_Finding Transformations.m`.

Bibliography

Antibiotic Resistance Threats in the United States (2013) Technical report, CDC—National Center for Health Statistics. https://www.cdc.gov/drugresistance/threat-report-2013/

Avinoam O, Schorb M, Beese CJ, Briggs JAG, Kaksonen M (2015) Endocytic sites mature by continuous bending and remodeling of the clathrin coat. Science (New York, NY) 348(6241):1369–1372. ISSN: 1095-9203. https://doi.org/0.1126/science.aaa9555

Curwin AJ, Brouwers N, Alonso Y Adell M, Teis D, Turacchio G, Parashuraman S, Ronchi P, Malhotra V (2016) ESCRT-III drives the final stages of CUPS maturation for unconventional protein secretion. eLife. ISSN: 2050-084X. https://doi.org/10.7554/eLife.16299

de Boer P, Hoogenboom JP, Giepmans BNG (2015) Correlated light and electron microscopy: ultrastructure lights up! Nat Methods 12(6):503–513. ISSN: 1548-7105. https://doi.org/10.1038/nmeth.3400

de Chaumont F, Dallongeville S, Chenouard N, Hervé N, Pop S, Provoost T, Meas-Yedid V, Pankajakshan P, Lecomte T, Le Montagner Y, Lagache T, Dufour A, Olivo-Marin J-C (2012) Icy: an open bioimage informatics platform for extended reproducible research. Nat Methods 9(7):690–696. ISSN: 1548-7105. https://doi.org/10.1038/nmeth.2075

Hampoelz B, Mackmull M-T, Machado P, Ronchi P, Huy Bui K, Schieber N, Santarella-Mellwig R, Necakov A, Andrés-Pons A, Philippe JM, Lecuit T, Schwab Y, Beck M (2016) Pre-assembled nuclear pores insert into the nuclear envelope during early development. Cell 166(3):664–678. ISSN: 1097-4172. https://doi.org/10.1016/j.cell.2016.06.015

Image Registration. https://www.mathworks.com/discovery/image-registration.html

Kukulski W, Schorb M, Welsch S, Picco A, Kaksonen M, Briggs JAG (2011) Correlated fluorescence and 3D electron microscopy with high sensitivity and spatial precision. J Cell Biol 192(1):111–119. ISSN: 1540-8140. https://doi.org/10.1083/jcb.201009037

Kukulski W, Schorb M, Kaksonen M, Briggs JAG (2012) Plasma membrane reshaping during endocytosis is revealed by time-resolved electron tomography. Cell 150(3):508–520. ISSN: 1097-4172. https://doi.org/10.1016/j.cell.2012.05.046

Paul-Gilloteaux P, Heiligenstein X, Belle M, Domart M-C, Larijani B, Collinson L, Raposo G, Salamero J (2017) eC-CLEM: flexible multidimensional registration software for correlative microscopies. Nat Methods 14(2):102–103. ISSN: 1548-7105. https://doi.org/10.1038/nmeth.4170

Schellenberger P, Kaufmann R, Siebert CA, Hagen C, Wodrich H, Grünewald K (2014) High-precision correlative fluorescence and electron cryo microscopy using two independent alignment markers. Ultramicroscopy 143:41–51. ISSN: 879-2723. https://doi.org/10.1016/j.ultramic.2013.10.011

Schindelin J, Arganda-Carreras I, Frise E, Kaynig V, Longair M, Pietzsch T, Preibisch S, Rueden C, Saalfeld S, Schmid B, Tinevez J-Y, White DJ, Hartenstein V, Eliceiri K, Tomancak P, Cardona A (2012) Fiji: an open-source platform for biological-image analysis. Nat Methods 9(7):676–682. ISSN: 1548-7105. https://doi.org/10.1038/nmeth.2019

Supplementary Information

K. Miura, N. Sladoje (eds.), *Bioimage Data Analysis Workflows*, Learning Materials in Biosciences,
https://doi.org/10.1007/978-3-030-22386-1

Index

www.ingramcontent.com/pod-product-compliance
Lightning Source LLC
LaVergne TN
LVHW082235200926
848484LV00008B/2357